Offshore Structural Engineering
Reliability and Risk Assessment

Offshore Structural Engineering
Reliability and Risk Assessment

Srinivasan Chandrasekaran

CRC Press
Taylor & Francis Group
Boca Raton London New York

CRC Press is an imprint of the
Taylor & Francis Group, an **informa** business

CRC Press
Taylor & Francis Group
6000 Broken Sound Parkway NW, Suite 300
Boca Raton, FL 33487-2742

First issued in hardback 2019

ISBN-13: 978-1-4987-6519-0 (hbk)

Library of Congress Cataloging-in-Publication Data

Names: Chandrasekaran, Srinivasan.
Title: Offshore structural engineering : reliability and risk assessment / Srinivasan Chandrasekaran.
Description: Boca Raton : Taylor & Francis, 2016. | "A CRC title." | Includes bibliographical references and index.
Identifiers: LCCN 2015045177 | ISBN 9781498765190 (alk. paper)
Subjects: LCSH: Offshore structures. | Structural engineering.
Classification: LCC TC1665 .C458 2016 | DDC 627/.98--dc23
LC record available at http://lccn.loc.gov/2015045177

Visit the Taylor & Francis Web site at
http://www.taylorandfrancis.com

and the CRC Press Web site at
http://www.crcpress.com

Contents

List of Figures xi
List of Tables xv
Preface xvii
Author xix

Chapter 1 Concept of Probability and Sampling Statistics 1

1.1 Introduction 1
1.2 Reliability and Risk 2
1.3 Types of Uncertainties 2
1.3.1 Parameter Uncertainties 3
1.3.2 Parametric Variability 3
1.3.3 Structural Uncertainties 3
1.3.4 Algorithmic Uncertainties 3
1.3.5 Experimental Uncertainties 4
1.3.6 Interpolation Uncertainties 4
1.4 Forward Uncertainty Propagation 4
1.5 Bayesian Approach 5
1.5.1 Modular Bayesian Approach 5
1.5.2 Full Bayesian Approach 6
1.6 Rules of Probability 7
1.7 Principles of Plausible Reasoning 8
1.8 Deductive Logic 9
1.9 Deductive Reasoning 13
1.9.1 Quantitative Rules 14
1.9.1.1 Product Rule 14
1.9.1.2 Sum Rule 17
1.9.2 Qualitative Rules 17
1.10 Continuous Probability Distribution Functions 19
1.11 Testing of Hypotheses 20
1.12 Simple and Compound Hypotheses 22
1.13 Urn Distribution 23
1.14 Random Variables 26
1.14.1 Generation of Random Numbers with a Specified Distribution 26
1.14.2 Multiple Random Variables 27
1.14.3 Nataf Random Variables 27
1.14.4 Random Variables Defined by Their Conditional Distributions 28
1.15 Monte Carlo Simulation Method 28
1.16 Importance of Sampling 30
1.17 Directional Simulation 34

1.18 Statistical Theories of Extremes 36
1.19 Modeling of Environmental Loads 37
1.19.1 Return Period 37
1.20 Estimate of Distribution Parameters 38
1.20.1 Method of Moments 39
1.20.2 Maximum Likelihood 39
1.20.3 Least Squares 39
1.20.4 Probability Plots 40
1.21 Exercise 40

Chapter 2 Structural Reliability Theory 59

2.1 Reliability 59
2.2 Variables in Reliability Study 61
2.3 Probabilistic Approach 62
2.4 Reliability Levels 63
2.5 Space of Variables 64
2.6 Error Estimation 64
2.6.1 Classification of Errors 65
2.7 Reliability and Quality Assurance 65
2.8 Uncertainties Inherent in Design 65
2.9 Uncertainties in System Design of Offshore Structures 67
2.10 Reliability Problem 68
2.11 Reliability Methods 69
2.12 First-Order Second Moment Method 69
2.13 Hasofer–Lind Method 71
2.14 Second-Order Reliability Methods 75
2.15 Simulation-Based Reliability Method 76
2.16 Reliability Estimate Using Higher-Order Response Surface Methods 76
2.17 High-Order Stochastic Response Surface Method 77
2.18 System Reliability 78
2.18.1 Series System 79
2.18.2 Parallel System 80
2.18.3 *k*-out-of-*n* System 80
2.19 General Systems 80
2.19.1 Cut Sets 80
2.19.2 Path Sets 81
2.20 System Functions for General Systems 81
2.21 Computing System Reliability 83
2.22 First-Order Estimates 86
Appendix A: Tutorials and Solutions 90
Appendix B: Application Problems 98
B.2.1 Mathieu's Stability 98
B.2.2 Stability of Tethers under Distinctly High Sea Waves and Seismic Excitation 105

Chapter 3 Reliability Analysis ... 119

3.1 Introduction ... 119
3.2 Fundamental Analysis ... 120
3.2.1 System with Equally Correlated Elements ... 120
3.2.2 System with Unequal Correlated Elements ... 122
3.3 Reliability Bounds for Structural Systems ... 124
3.4 Application of Structural Codes on Safety ... 125
3.5 Limit State Functions ... 126
3.6 Characteristic Value of Basic Variables ... 126
3.6.1 Treatment of Geometric Variables ... 126
3.6.2 Treatment of Material Properties ... 126
3.6.3 Treatment of Load and Other Actions ... 126
3.6.4 Evaluations of Partial Coefficient ... 127
3.7 Stochastic Modeling ... 127
3.8 Mechanical Modeling ... 129
3.9 Mechanical Model and Reliability Coupling ... 131
3.10 Complexity of Mechanical Model and Reliability Coupling ... 132
3.10.1 Complexities in Geometric Modeling of Hinged Joints ... 133
3.11 Stochastic Process ... 135
3.12 Gaussian Process ... 138
3.13 Barrier Crossing ... 139
3.14 Peak Distribution ... 139
3.15 Fatigue Reliability ... 140
3.15.1 Discrete Wave and Spectral Methods of Fatigue Analysis ... 140
3.16 *S–N* Curve and Fatigue Damage ... 141
3.17 Estimate of Cumulative Damage (Linear Damage Hypothesis) ... 144
3.18 Design *S–N* Curves ... 144
3.19 Fatigue Assessment Using Discrete Wave Approach ... 145
3.19.1 *S–H* Relationship ... 145
3.19.2 Fatigue Damage ... 146
3.20 Simplified Fatigue Assessment Method ... 148
3.20.1 Effect of Dynamic Amplification ... 148
3.21 Spectral Fatigue Analysis of Offshore Structures ... 149
3.22 Short-Term Fatigue Damage ... 150
3.22.1 Evaluation of Damage Integrals ... 151
3.23 Uncertainties in Fatigue Reliability ... 152
3.24 Lognormal Format for Fatigue Reliability ... 153
3.25 Tubular Joints: Experimental and Analytical Investigations ... 154
3.25.1 Fatigue Life Estimate of Tubular Joints ... 156
3.26 Behavior of T Joints under Axial Loads ... 157

3.27 T Joint under Out-of-Plane Bending 166
3.28 K Joints under Axial Loading 171

Chapter 4 Risk Assessment 177

4.1 Introduction 177
4.2 Quantified Risk Assessment 178
4.3 Hazard Identification 178
4.4 Hazard and Operability 178
4.4.1 Applicability 179
4.5 HaZop Study Process 180
4.5.1 Node Identification 180
4.5.2 Operating Modes 181
4.5.3 Lifecycle Changes 181
4.6 Parameters for HaZop Study 181
4.6.1 Guide Words 182
4.6.2 Deviations 183
4.6.3 Causes 183
4.6.4 Consequences 184
4.6.5 Safeguards 184
4.7 HaZop: Advantages and Limitations 187
4.7.1 Limitations of HaZop 187
4.7.2 Example Illustration 187
4.8 Logical Risk Analysis 188
4.8.1 Risk Index 189
4.8.2 Relative Risk 193
4.8.3 Example Problem for Risk Analysis 194
4.9 Failure Mode and Effect Analysis 195
4.9.1 FMEA Methodology 195
4.9.2 FMEA Applications 197
4.9.3 FMEA Variables 198
4.10 Fault Tree and Event Tree 199
4.11 Fault Tree Analysis 201
4.12 Event Tree Analysis 203
4.13 Cause–Consequence Analysis 203
4.14 Decision Trees 204
4.15 Consequence Analysis 205
4.15.1 Source-Strength Parameters 205
4.15.2 Consequential Effects 205
4.16 Limitations of QRA 206
4.17 Risk Acceptance Criteria 207
4.18 Risk and Hazard Assessment 208
4.19 Hazard Identification 209
4.20 Selection of Failure Scenarios 210
4.20.1 Blocking Systems 211
4.20.2 Excess Flow Valve 212

4.20.3 Non-Return Valve 212
4.20.4 Bund 212
4.20.5 Intervention by Operators 212
4.21 Fire and Thermal Radiation 213
4.21.1 Jet Fires 213
4.21.2 Pool Fires 213
4.21.3 Fireball or Boiling Liquid Expanding Vapor Explosion 213
4.21.4 Vapor Cloud Explosion 213
4.22 Selection of Damage Criteria 214
4.22.1 Heat Radiation 214
4.22.2 Explosion 215
4.23 Risk Picture 216
4.24 Individual Risk 216
4.25 Societal Risk 217
4.26 Risk Assessment and Management 218
4.26.1 Objectives of Risk Management 218
4.27 Example Problem of Risk Assessment: Offshore Triceratops 220

Model Exercise Papers 227

References 239

Index 247

List of Figures

Figure 1.1 Plot of the variables 27

Figure 1.2 Sampling within hyper-rectangle 31

Figure 1.3 Improvement of failure estimate in two-dimensional domain 32

Figure 1.4 Typical case of failure domain 33

Figure 1.5 Sampling with normal density at design point 33

Figure 1.6 Failure boundary with CDF of chi-square distribution 35

Figure 2.1 Components of reliability analysis 60

Figure 2.2 Interrelationship of variables in reliability study 62

Figure 2.3 Relation between reliability index and probability of failure 63

Figure 2.4 Space of variables marking safe and failure domains 64

Figure 2.5 Concept of reliability with two random variables 68

Figure 2.6 Time-variant reliability problem 69

Figure 2.7 Hasofer–Lind reliability index with linear performance function 72

Figure 2.8 Hasofer–Lind reliability index with nonlinear performance function 74

Figure 2.9 Parallel system consisting of *n* components 80

Figure 2.10 A general system 81

Figure 2.11 General system with minimum cut sets 82

Figure 2.12 General system with minimum path sets 82

Figure 2.13 Failure domains for a series system consisting of three components 84

Figure 2.14 Failure domains for a parallel system consisting of three components 84

Figure 2.15 Failure domain for the general system 85

Figure 2.16 System failure domain of first-order approximation 89

Figure A2.1 cdf and pdf for the example problem. CCDF—Complementary cumulative distribution function 95

Figure B2.1 Plan of TLP, considered for the example problem 101

Figure B2.2 Mathieu's stability chart 102

Figure B2.3 Generalized displacement amplitude for TLP_1 tethers—first mode ($C_m = 1.5$) ... 103

Figure B2.4 Generalized displacement amplitude for TLP_1 tethers—first mode ($C_m = 2.0$) ... 103

Figure B2.5 Generalized displacement amplitude for TLP_2 tethers—first mode ($C_m = 1.5$) ... 103

Figure B2.6 Generalized displacement amplitude for TLP_2 tethers—first mode ($C_m = 2.0$) ... 104

Figure B2.7 Generalized displacement amplitude for Auger TLP tethers—first mode ($C_m = 1.5$) ... 104

Figure B2.8 Generalized displacement amplitude for Auger TLP tethers—first mode ($C_m = 2.0$) ... 105

Figure B2.9 PM spectrum and sample wave history used in the study ... 108

Figure B2.10 Example TLP taken for the study ... 109

Figure B2.11 PSD of the vertical ground acceleration ... 111

Figure B2.12 Time history of ground velocity ... 112

Figure B2.13 Response time histories of triangular TLPs under seismic forces and high sea waves ... 115

Figure B2.14 Frequency domain response of TLPs under seismic forces and high sea waves ... 116

Figure B2.15 Phase plots of response of TLPs under seismic and high sea waves ... 117

Figure 3.1 Representations of variables ... 127

Figure 3.2 Gaussian density plot ... 128

Figure 3.3 Approximated representation of physical reality ... 130

Figure 3.4 Failure scenarios in mechanics ... 130

Figure 3.5 Reliability sensitivity analyses ... 131

Figure 3.6 Reliability analysis ... 132

Figure 3.7 Coupling of mechanical model and reliability coupling ... 132

Figure 3.8 Offshore triceratops ... 134

Figure 3.9 Hinged joint connecting deck and buoyant legs ... 134

Figure 3.10 Schematic diagram to find stiffness and damping of the hinged joint ... 135

Figure 3.11 Rotational stiffness of hinged joint in rotational degrees of freedom (scaled values) ... 136

Figure 3.12 Realization of the stochastic variable ... 136

Figure 3.13 Realization and the corresponding density function ... 136

Figure 3.14 Realization of stochastic process ... 139

Figure 3.15 Two-segment *S–N* curve for different stress ranges ... 142

Figure 3.16 Nomenclature of the joint ... 155

Figure 3.17 T joint and K joint fabricated for the study ... 156

Figure 3.18 Geometric details of (a) T joint and (a) K joint considered for the study ... 158

Figure 3.19 Experimental setup with components ... 158

Figure 3.20 Sample fixed with strain rosette ... 159

Figure 3.21 T joint under axial load set up ... 159

Figure 3.22 HSS variation across the profile for tubular joint under axial loading ... 160

Figure 3.23 SCF variation of T joint under axial loading ... 161

Figure 3.24 SCF variation around the T joint under axial load: experimental observations ... 162

Figure 3.25 SCF variation around T joint under axial load: numerical analyses ... 162

Figure 3.26 Comparison of SCF at T joint chord member ... 163

Figure 3.27 Comparison of SCF at T joint brace member ... 163

Figure 3.28 Failed T joint in punching shear ... 165

Figure 3.29 Failed T joint showing bulging of joint ... 165

Figure 3.30 Region of influence of T joint under axial load failure ... 165

Figure 3.31 T joint under out-of-plane bending ... 167

Figure 3.32 Hot-spot stress of T joint under out-of-plane bending ... 167

Figure 3.33 Hot-spot stress variation around the T joint under out-of-plane bending ... 167

Figure 3.34 SCF of T joint under out-of-plane bending (experimental) ... 168

Figure 3.35 Hot-spot stress contour in tubular T joint under out-of-plane bending ... 168

Figure 3.36 Stress concentration factor around weld of T joint under out-of-plane bending (numerical) ... 169

Figure 3.37 Comparison of SCF at T joint chord member ... 169

Figure 3.38 Comparison of SCF at T joint brace member 170
Figure 3.39 Failure mode of tubular T joint under out-of-plane bending 172
Figure 3.40 K joint under axial load 173
Figure 3.41 SCF variation around chord of K joint under balanced axial load 173
Figure 3.42 SCF variation around chord of K joint under out-of-plane bending 174
Figure 3.43 SCF of K joint under axial load (numerical analysis) 175
Figure 4.1 HaZop study process 180
Figure 4.2 Isolated segment of process line for HaZop study 188
Figure 4.3 FMEA road map 198
Figure 4.4 Logical gates for FTA 201
Figure 4.5 Symbols used in FTA 201
Figure 4.6 Typical combination of logical gates of FTA 202
Figure 4.7 Block diagram for system reliability 202
Figure 4.8 Event tree 203
Figure 4.9 Cause–consequence chart 204
Figure 4.10 Decision tree for risk assessment 204
Figure 4.11 ETA for liquid release 206
Figure 4.12 Risk acceptance criteria 207
Figure 4.13 Flowchart of QRA 209
Figure 4.14 Risk assessment and management 218
Figure 4.15 Risk assessment matrix 219
Figure 4.16 Offshore triceratops 219
Figure 4.17 FMEA methodology used for example problem 220
Figure 4.18 Component breakdown and failure modes of offshore triceratops 221

List of Tables

Table 2.1 Practical and Theoretical Reliability....60

Table 2.2 Geometric Properties of Tension Leg for Stability Analysis....102

Table 2.3 Stability Parameters for TLP_1....102

Table 2.4 Stability Parameters for TLP_2....102

Table 2.5 Stability Parameter for Auger TLP....105

Table B2.1 Geometrical Properties of Example TLPs (Square)....112

Table B2.2 Coefficient for Interpolation of C_m....113

Table B2.3 Natural Wave Periods and Natural Frequencies of Triangular TLPs....113

Table B2.4 Response Summary of Triangular TLPs....114

Table 3.1 Parameters for Class "T" Offshore *S–N* Curves....143

Table 3.2 Parameters for Plate Thickness Adjustment....143

Table 3.3 Details of Parameters of T Joint under Axial Loading....160

Table 3.4 SCF Equation and Parameter Validation for T Joint under Axial Loading....160

Table 3.5 Constant k_1 Value....164

Table 3.6 Allowable Number of Cycles for T Joint for Different SCF Values....164

Table 3.7 Fatigue Life of T Joint under Axial Loading....164

Table 3.8 SCF Variation around T Joint Weld by a Different Method....166

Table 3.9 T Joint under Out-of-Plane Bending....166

Table 3.10 SCF Equation and Parameter Validation for T Joint under OPB Loading....168

Table 3.11 SCF for T Joint under Out-of-Plane Bending (Parametric)....169

Table 3.12 Constant k_1 for T Joint under Out-of-Plane Bending....170

Table 3.13 Allowable Number of Cycles for T Joint under Out-of-Plane Bending Using Experimental and Parametrical SCF Values....170

Table 3.14 Fatigue Life of T Joint under Out-of-Plane Bending....171

Table 3.15 Variations in Stress Concentration Factor around T Joint Weld under Out-of-Plane Bending....172

Table 3.16 K Joint under Axial Load ... 173

Table 3.17 SCF Equation and Parameter Validation for K Joint under Axial Load ... 174

Table 3.18 SCF for K Joint under Axial Load (Experimental Investigations) ... 174

Table 3.19 SCF for K Joint under Axial Load (Parametric) ... 175

Table 3.20 Constant k_1 for K Joint under Axial Load ... 175

Table 3.21 Allowable Number of Cycles for K Joint under Axial Load Using Experimental and Parametrical SCF Values ... 175

Table 3.22 Fatigue Life of K Joint under Axial Load ... 176

Table 3.23 Variations in Stress Concentration Factor for K-Joint Weld under Axial Load ... 176

Table 4.1 Parameter and Guide Word Combinations for HaZop Study ... 183

Table 4.2 Typical HaZop Record Sheet ... 186

Table 4.3 HaZop Sheet for the Example Problem ... 188

Table 4.4 Parameters Useful for Risk Analysis ... 190

Table 4.5 Format of Reporting FMEA Study ... 197

Table 4.6 FMEA Worksheet, Modified with Details ... 200

Table 4.7 Acceptable Risk Criteria ... 208

Table 4.8 Probabilities of Failure of Excess Flow Valve ... 212

Table 4.9 Effect of Heat Radiation on Humans and Equipment ... 214

Table 4.10 Damage to Human Life with Respect to Time of Exposure ... 215

Table 4.11 Effect of Overpressure ... 215

Table 4.12 Risk Analysis Summary for Offshore Triceratops ... 222

Table 4.13 FMEA-Offshore Triceratops ... 223

Preface

Offshore structures, built for extraction of various natural resources, have evolved over many decades in their design concepts. Today, design of offshore facilities needs to address various uncertainties involved in different stages, including data collection, analysis, design and construction, and commissioning. A detailed knowledge of the reliability of offshore structures using probabilistic tools becomes a necessity for both industry and academia. Risk is always present in human activity and there are failure modes that one must identify carefully. Quality assurance is the first condition of reliability though it is always possible to take refuge behind the rules for innovating new offshore structures. "Too strong to fail" can prove to be expensive in the present climate and can even be inefficient. An innovative design must be a reliable design, which is the objective of this book's content.

Risk assessment and reliability estimates are generally based on probability. It is usual practice that books on probability do not approach the numbers with the reliability window and vice versa, although there exists a very strong bond between them. A chapter on probability theory including plausible reasoning is included in an attempt to bridge this gap. In simple terms, this is user-friendly text material that will facilitate self-learning for a course on risk and reliability. Applications, with respect to offshore structures, which are one of the highest uncertainties in engineering structures, are addressed. Not all the contents of this book will be brand new but the approach to their explanation certainly is. The book is a good teaching guide for faculty in engineering schools and practicing professionals. As offshore structures are one of the most expensive types of infrastructures, it is imperative to estimate the risk of the structure and its reliability index. As reliability estimates of offshore installations are becoming mandatory under revised insurance legislations, the subjects covered in this book are therefore of societal importance and interest to the engineering community.

I sincerely thank Indira Raghavan for editorial assistance extended during the preparation of the manuscript. The technical assistance by research scholars Lognath Seeramulu, Shihas Khader, Venkatakiran, and Rohit Srivastava is worth mentioning and I sincerely acknowledge their help and assistance.

Thanks are due to all authors who dealt with this subject in the past and from whom I developed the courage to document my understanding in a book. I also acknowledge major contributions made by the relatively small number of researchers in the field of structural reliability. I thank the Centre for Continuing Education, IIT Madras, India for their administrative support extended during the manuscript preparation. Additional support in the form of a power-point presentation is offered (http://www.crcpress.com/product/isbn/9781498765190), which may be useful in following the book as a detailed course for graduate level teaching at universities. The administrative authorities and editorial team of CRC Press deserve special mention for their excellent professional guidance in bringing this book to its present form.

Srinivasan Chandrasekaran

Author

Srinivasan Chandrasekaran is a professor in the Department of Ocean Engineering, Indian Institute of Technology Madras, India. He has more than 25 years of teaching, research, and industrial experience and has supervised many sponsored research projects and offshore consultancy assignments in India and abroad. His active areas of research include dynamic analysis and design of offshore platforms, development of geometric forms of compliant offshore structures for ultra-deep water oil exploration and production, subsea engineering, rehabilitation and retrofitting of offshore platforms, structural health monitoring of ocean structures, seismic analysis and design of structures, and risk analyses and reliability studies of offshore and petroleum engineering plants. For a period of 2 years, Dr. Chandrasekaran was a visiting fellow by invitation of the Ministry of Italian University Research to the University of Naples Federico II, Italy. During this time he conducted research on advanced nonlinear modeling and analysis of structures under different environmental loads with experimental verifications. He has published about 140 research papers in international journals and refereed conferences organized by professional societies around the world. Dr. Chandrasekaran is a member of many national and international professional bodies and has delivered many invited lectures and keynote addresses at international conferences, workshops, and seminars organized in India and abroad.

1 Concept of Probability and Sampling Statistics

1.1 INTRODUCTION

Construction of offshore structures targets deep and ultra-deep waters in search of oil and gas. In general, the structural forms of offshore structural systems are designed to remain flexible as they can alleviate the encountered environmental loads effectively; they are also required to remain robust and safe at all levels of operations (Srinivasan Chandrasekaran, 2014a,b,c,d, 2015a,b). New and innovative materials that are increasingly being used in the construction of offshore structures demand more detailed understanding of their mechanical properties under critical stages of functioning (Srinivasan Chandrasekaran, 2015c). While environmental loads that act on offshore structures are also complex in their action, their estimates also have many uncertainties (Tromans et al., 1992). Environmental loads are not completely known but only some of their features are known; few examples are wave loads, wind loads, ice loads in arctic regions, seismic loading on subsea systems, etc. Randomness in the environmental loading also increases these uncertainties during the planning, analysis, and design and construction stages. At all these stages (i) accuracy in estimating the above loads; and (ii) construction process and techniques used also add to a different level of complexities. Reliability tools are seen to be effective in handling this wide range of uncertainties (Assakkaf and Ayyub, 2000; Augusti et al., 1984).

Reliability methods have many advantages namely: (i) offering a realistic possession of uncertainties; (ii) offering methods to evaluate safety factors, which are otherwise chosen arbitrarily; (iii) offering decision-making support for more economic and better-balanced design; (iv) enabling the analysis of different modes of failure and measuring the reliability provided by the application of code regulations; (v) allowing an optimal use of materials amongst various components of the structure; and (vi) enabling the expansion of the knowledge of uncertainties in the response of structures (Bierager, 1990; Datta, 2010; Ditlevsen and Madsen, 1996; Ditlevsen, 1981; DNV, 1992). However, there exist a few limitations namely: (i) probabilities (to a major extent) and statistics (to a lesser extent) are more a part of the knowledge of the mathematician who can doubt, rather than that of the engineer who must have certainties; and (ii) explicitly underscoring the acceptance of a risk, which is only an implicit (John Spouge, 2005; Kiran, 2012). Even if it is justified by using safety coefficients, it still needs an engineering-based judgment to support the decision process (Melchers, 1999, 1987). In general, reliability methods project the overall process of analysis, design, and construction of offshore structures as a probable solution and not exact ones. Hence, decisions, supported by experience are always questioned for a better alternative (Madsen et al., 1986; Malhotra and Penzien, 1970).

While uncertainties need to be expressed in probabilistic terms for their better representation, failure is also then expressed in terms of probability. This is because there is some degree-of-uncertainty in confirming the failure of any structure, as explained above. In general, failure is assessed by inability of the structure to perform its intended function adequately on demand for a specified period; it is very important to note that the assurance is true only under *specific conditions*. Conversely, reliability is expressed in terms of success of a system to sustain the demand expected from it. Reliability is also expressed in terms of probability indicating the following vital parameters namely: (i) quality of performance; (ii) over an expected period; and (iii) to perform under *specific conditions*. This is the genesis of reliability, being implied as the probability of success to perform the intended function. Reliability implies the estimate of limit state probabilities of a structure under critical demand. Safety is generally used to indicate reliability, which is more a traditional concept.

1.2 RELIABILITY AND RISK

Reliability is defined as probability that a facility will perform its intended function for a specific period under defined conditions. It is expressed as the converse of failure (as $R = 1-P_f$), where P_f is the probability of failure. Risk is defined as a measure of magnitude of a hazard. The two constituents of risk are probability of failure (P_f) and frequency of failure (h_f) of a damaging event, E. Risk assessment deals with two fundamental questions namely: (i) what could happen in what way and how often?; (ii) what may be allowed to happen, how often and where? Answering the above two questions will lead to risk assessment (Throft-Christensen and Baker, 1982). For an offshore structural engineer, if the assessment turns out to be negative, further questions need to be answered such as what suitable measures are needed to provide the required safety? Then the engineer has to ensure that appropriate measures are put in place to guarantee proper functioning. Reliability is focused on those problem areas that are not realized directly by society but those that challenge safety indirectly (Srinivasan Chandrasekaran, 2015b, c). It is interesting to know that risk assessment from failure analyses of structures does not generate discussions but only lead to rules and recommendations. However, reliability opens into an engineering questionnaire that leads to a better understanding of (probability of) failure (Technip, 1981).

1.3 TYPES OF UNCERTAINTIES

As is seen and understood from the above discussions that uncertainties govern the use of reliability tools to judge the safe performance of offshore structures, there are different types of uncertainties that need to be known, namely: (i) those arising from the parameters; (ii) variables that are vital input for analysis and design; (iii) those arising from mathematical modeling; (iv) those arising from the mathematic algorithms being used in the analysis and design; (v) those arising during experimental investigations that are carried out to calibrate the response parameters of the chosen structural geometry; and (vi) those arising from interpolating the response parameters of a prototype from that of the experimental or numerical simulations (Trevor Kletz, 2003).

1.3.1 Parameter Uncertainties

They arise from the input to the mathematical model for analysis and design. They remain uncertain as to their exact values due to many reasons namely: (i) they are unknown to experimentalists; (ii) they cannot be controlled during the experimental investigations; and (iii) their values cannot be exactly inferred by any statistical methods. A few examples are damping estimates in offshore structures, compatibility behavior at connections (joints), coupling effect of different degrees of freedom in their real extent, effect of *P–M* interaction of universal joints (see, e.g., offshore triceratops), etc.

1.3.2 Parametric Variability

They arise from the input variables of the physical or numeric model. For example, limitations in the number of finite element mesh, cross-sectional dimensions of ball joints in hinged connections, tether tension variations that arise from the seismic activities at the seabed, etc., may not be exactly modeled as designed, which would cause variability in its performance.

1.3.3 Structural Uncertainties

They arise from the inaccuracy of mathematical models that simulate the real-time behavior of offshore structures under installed conditions (Srinivasan Chandrasekaran, 2015d). It is a known fact that numerical models approximate reality and cannot be modeled accurately. For example, it is difficult to model the behavior of pinned connections, as required in articulated towers under the combined action of axial load and moment. In addition, damping that arises from the geometric interferences of members in the top side that can cause serious variations in wind load estimates are difficult to model. In such cases, even if there are unknown parameters in the model, a discrepancy is still expected between the model and the true physics.

1.3.4 Algorithmic Uncertainties

They arise from the approximations per implementation of the analytical models. As the solution of equations of the motion of offshore structures is generally iterative, numerical methods are used to solve the system of equations; most models are too complicated to solve exactly. For example, the finite element method or finite difference method may be used to approximate the solution of a partial differential equation, which, however, introduces numerical errors. Numerical integration methods inherently deal with the infinite sum truncation that is necessary to approximations in the scheme of numerical implementation.

1.3.5 Experimental Uncertainties

They arise from the variability of experimental measurements. Experimental investigations, which are an inevitable part of the analysis of offshore structures, are generally circumscribed by serious limitations that arise from electronic sensors. Experimental uncertainties are inevitable and can be easily realized by repeating a set of measurements many times using exactly the same settings for all inputs/variables.

1.3.6 Interpolation Uncertainties

They arise from the lack of available data collected from computer model simulations and/or experimental measurements. For other input settings that do not have simulation data or experimental measurements, one must interpolate or extrapolate in order to predict the corresponding responses. The problem is more serious if the physical model of the structure does not represent the appropriate mass of the prototype for simulating the real dynamic behavior.

1.4 FORWARD UNCERTAINTY PROPAGATION

As discussed above, uncertainties are primarily responsible for making reliability studies probabilistic in nature. It is also important to note that it is the probability tag associated with reliability that makes it more mathematical; otherwise reliability is visualized as a study based on engineering judgment (Srinivasan Chandrasekaran, 2014a,b,c,d). As seen above, various types of uncertainties that are inherently present in the system (and cannot be avoided) need to be handled in the analysis and design. Existing uncertainty propagation includes both probabilistic and non-probabilistic approaches.

Probabilistic approaches for uncertainty propagation are of the following categories:

- *Simulation-based methods*: Monte Carlo simulation, importance sampling, adaptive sampling, etc.
- *Local expansion-based methods*: Taylor series, perturbation method, etc. These methods have advantages when dealing with relatively small input variability and outputs that do not express high nonlinearity.
- *Functional expansion-based methods*: Newman expansion, orthogonal or Karhunen–Loeve expansions (KLE), polynomial chaos expansion (PCE), and wavelet expansions are classical examples under this category.
- *Most probable point (MPP)-based methods*: First-order reliability method (FORM) and second-order reliability method (SORM).
- *Numerical integration-based methods*: Full factorial numerical integration (FFNI) and dimension reduction (DR).

Interval analysis, fuzzy theory, possibility theory, and evidence theory are among the most widely used techniques in non-probabilistic based approaches.

The probabilistic approach is considered as the most rigorous approach to uncertainty analysis in engineering design due to its consistency with the theory of decision analysis. This can be performed rigorously for random variables that are obtainable as transformations of Gaussian variables, leading to exact confidence intervals. In regression analysis and least squares problems, the standard error of parameter estimates is readily available, which can be expanded into a confidence interval.

1.5 BAYESIAN APPROACH

Several methodologies for inverse uncertainty quantification exist under the Bayesian framework; the objective of solving the associated problem with both bias correction and parameter calibration is the most complicated one. The challenges of such problems include not only the influences from model inadequacy and parameter uncertainty, but also the lack of data from both numerical simulations and experimental investigations (Terje Aven and Jan Erik, 2007).

1.5.1 Modular Bayesian Approach

An approach to inverse uncertainty quantification is the modular Bayesian approach (Ang and Tang, 1984, 1975; Blight and Ott, 1975; Breitung, 1984). The modular Bayesian approach derives its name from its four-module procedure. Apart from the current available data, a prior distribution of unknown parameters should be assigned.

Module 1 deals with the Gaussian Process (GP) modeling for the numerical model. To address the issue from the lack of simulation results, the numerical model is replaced with a GP model. This is given by

$$y^m(X,\theta) \sim \mathrm{gp}(h^m(.)^T \beta^m, \sigma_m^2 R^m(.,.)) \tag{1.1}$$

$$R^m\left((X,\theta),(X',\theta')\right) = \exp\left\{-\sum_{K=1}^{d} \omega_K^m (x_k - x_K')^2\right\} \exp\left\{\sum_{K=1}^{r} \omega_{d+K}^m (\theta_K - \theta_K')^2\right\} \tag{1.2}$$

where d, which is the limit of summation, indicates the dimension of the input variables and r is that of the unknown parameters. While $h^m(\cdot)$ is predefined, $\{\beta^m, \sigma_m, \omega_K^m, K = 1,\ldots,d+1\}$ are known as *hyper parameters* of the GP model, which need to be estimated through the maximum likelihood estimate (MLE). This module can be considered as a generalized Kriging method (Box and Tiao, 1973).

Module 2 deals with the GP modeling for the discrepancy function. Similar to the first module, the discrepancy function is replaced with a GP model, as given below:

$$\delta(X) \sim \mathrm{gp}(h^\delta(.)^T \beta^\delta \sigma_\delta^2 R^\delta(.,.)) \tag{1.3}$$

$$R^{\delta}(X,X) = \exp\left\{-\sum_{K=1}^{d} \omega_K^{\delta}(x_K - x_K')^2\right\} \tag{1.4}$$

In this case, along with the prior distribution of unknown parameters and data obtained from both the numerical models and experiments, one can derive the maximum likelihood estimates (MLEs).

$$\{\beta^{\delta}, \sigma_{\delta}, \omega_k^{\delta}, K = 1,\ldots,d\} \tag{1.5}$$

At the same time, β^m from Module 1 gets updated as well.

Module 3 deals with the posterior distribution of unknown parameters. Bayes' theorem is applied to calculate the posterior distribution of the unknown parameters as given below:

$$p(\theta \mid data, \Phi)\ \alpha\ p(data \mid \theta, \Phi)p(\theta) \tag{1.6}$$

where Φ includes hyper parameters of both the previous modules.

Module 4 deals with the prediction of the experimental response and discrepancy function.

1.5.2 Full Bayesian Approach

The full Bayesian approach requires the prior estimate of unknown parameters θ and hyper parameters Φ as well. The following steps are important:

1. Derive the posterior distribution $p(\theta, \Phi \mid data)$.
2. Integrate Φ out and obtain $p(\theta \mid data)$. This step accomplishes the calibration.
3. Prediction of the experimental response and discrepancy function.

However, this approach has significant drawbacks as listed below:

- For most cases, $p(\theta \mid data)$ is a highly intractable function of Φ. Hence the integration becomes complex. Moreover, if priors for the other hyper parameters Φ are not carefully chosen, complexity in the numerical integration increases even more.
- In the prediction stage, the prediction (which should at least include the expected value of system responses) also requires numerical integration. The Markov chain Monte Carlo (MCMC) is often used for integration; however it is computationally expensive.

It is interesting to note that the full Bayesian approach is computationally expensive and may not yet be practical to deal most complicated modeling situations, as required in offshore structural analyses (Bucher and Bourgund, 1990; Bury, 1984).

1.6 RULES OF PROBABILITY

The following important rules of probability that are frequently employed in reliability studies are derived from the understanding of plausible reasoning:

Rule 1:

Let E be the event to occur. $P[E]$ is a simple number to express the confidence in the occurrence of that event. Thus $P[E]$ is the probability that translates mathematically and consistently the confidence in occurrence of event E. This is called subjective probability.

Rule 2:

Let E be the event of interest.

$$P[E] \neq P[E \mid H] \tag{1.7}$$

This statement implies that probability of event E is now different from that of the earlier ones. It is due to the fact that with the given hypothesis (H), the knowledge status changes the probability of the event, E. However, it is always conditional to the given hypothesis.

Rule 3:

In estimating $P[E]$, there can be other alternatives. All the alternatives put together is called a *space of events*, S. $P[S] = 1$; this is a union of all possible alternative events. Hence, the following statement is valid:

$$S = \text{U}\{E\text{U}C\text{U}S\text{U}F\} \text{ and } P[S] = 1 \tag{1.8}$$

Rule 4:

If we now group the alternative events as "either E can occur or everything else can occur," then the following statement holds good:

$$P[E\text{U}A] = P[E] + P[A] \tag{1.9}$$

The above statement is valid under the condition that A and E are mutually exclusive. If they are not mutually exclusive, then the following statement is valid:

$$P[E\text{U}A] \neq P[E] + P[A] \tag{1.10}$$

Rule 5:

For a given event E, we assign either of the following:

$$P[E] = 0 \text{ when for sure, event } E \text{ will not occur} \tag{1.11}$$

$$P[E] = 1 \text{ when for sure, } E \text{ will occur} \tag{1.12}$$

$$P[E] \in [0.1] \text{implies that } E \text{ can happen, probably}$$

Rule 6:

$$P(E \mid C) = \frac{P[E \cap C]}{P[C]} \tag{1.13}$$

Rule 7:

$$P[E \cap C] = P[E \mid C] * P[C] = P[C \mid E] * P[E] \tag{1.14}$$

Rule 8:

$$P[E \cup C] = P[E] + P[C] - P[E \cap C] \tag{1.15}$$

Rule 9:

If $P[E \mid C]$ is known and $P[E \mid \bar{A}]$ is also known, $P[E]$ is given by the following relationship:

$$P[E] = P[E \mid C] * P[C] + P[E \mid \bar{C}] * P[\bar{C}] \tag{1.16}$$

This is called the Total Probability Theorem.

Rule 10:

Given event E, what is the probability that this event occurs due to the occurrence of a specific event I? This is a reverse problem and can be estimated using Bayes' theorem as given below:

$$P[I \mid E] = \frac{P[E \cap I]}{P[E]} \tag{1.17}$$

$$= \frac{P[E \mid I] * P[I]}{P[E]} \tag{1.18}$$

$$P[I \mid E] = \frac{P[E \mid I] * P[I]}{P[E \mid I] * P[I] + P[E \mid I] * P[I]} \tag{1.19}$$

1.7 PRINCIPLES OF PLAUSIBLE REASONING

Rules of thinking are generally termed as plausible reasoning. Probability theories are complex in nature and one finds it difficult to learn as they try to model everything that a human brain thinks (Benjamin and Cornell, 1970). However, the system for probability, based on plausible reasoning, looks abstract in the beginning but everything becomes derivable. This leaves no chance for confusion. Probability theory is nothing but common sense reduced to calculation. Probabilistic analysis has two classes namely: (i) Bayesian class; and (ii) frequency class. In the Bayesian class, prior information of the subject is included, which improves the knowledge status of the unknown (Barlow and Prroschan, 1981). Derived information is based on the available data, called posterior information. Frequency class is based on the sampling distributions. This is not capable of incorporating the prior information. It further assumes that all the realizations with the sample are independent.

Plausible reasoning can be explained with a simple example. Consider the following event. On a late night, policemen were patrolling on a main road of a city center. They heard a loud sound similar to that of breaking glass. They also noticed a man running in the dark wearing a mask on his face. The man was also carrying a big bag in his hand and the policemen could think that he was a thief. The incident also has other similar options as follows:

Policemen did not know complete information about the incident. The person who was noticed to be running in the dark could also be the owner of the store who was coming out of the shop. Instantly, there was a loud sound being heard, which was similar to that of breaking glass. However, the fact remains verified that the shop window was found broken, which must have been damaged earlier. The man who was walking by the side of the shop with a bag was considered to be a thief. While the policemen decided that he was a thief, he could also be the owner of the shop, who was returning home after closing the shop. Now, let us examine the reasons for the policemen thinking that the man was a thief. One of the main reasons could be due to the past (similar) experience, it is plausible to think of the man as the thief. Therefore, the following propositions hold true:

A: Window glass broken; man with mask; sneaking out; having a big bag in his hand; dark night
B: Man is the thief

Given B is true $\Rightarrow$ A is more plausible—this is a direct problem.

What the policemen see is assigned as A, which is true; if A is seen as true, they decide that the man is a thief. In fact, if A is seen as true, B becomes more plausible. This is an inverse problem. The following logic is interesting and highly relevant in plausible reasoning.

1.8 DEDUCTIVE LOGIC

If A is true, then B is true: This is a hypothesis. Considering the example cited above, it is seen that the policemen observed that A is true. This is established, which means that B is true. This is the famous $A \Rightarrow B$ statement. This also means that if B is false, A is false. This statement confirms that there is only a logical dependence and no physical dependence. Let us consider another example to elaborate upon this logic. Let event A be a statement that "It is going to rain at 12:30" and B represent an event that "clouds became dark before 12:30." If A is true, then B is true, which means that $A \Rightarrow B$ (A implies B) or if B is false, then A must be also false. In this logic, A and B has only two values, namely true or false. Applying this to offshore engineering, let A be an event representing "loss of strength and durability" and B represent "steel is corroded." The above logic fits very well in the present case, which is useful in establishing the reasons for loss of strength and factors that affect durability. But, more interestingly, such deductive logic that has a straightforward and strong reasoning is not more practical; in plausible reasoning, one is more interested in the weaker reasoning. If A is true, then B is true. This means that $A \Rightarrow B$. this is called deductive logic. For B becomes true, A becomes more plausible. If A is false, then B becomes less plausible. For example, at 12:25, there is no dark cloud, then rain at 12:30 becomes less plausible. Therefore, it is important to note that in the case of deductive reasoning, conclusions have certainty. Thus, plausible reasoning shall be helpful in arriving conclusions through probability theories.

Let us now relook at the policeman example again. Given A was true, the policemen made up their minds to conclude that the man was a thief. This is based on similar experience they had in the past. Kindly note that plausible reasoning makes the conclusions more justifiable; this is true when the knowledge status about A is improved. Further, let us try to understand the deductive logic more in detail. $A \Rightarrow B$ means that if A is true, then B is true; if B is false, A must be false. A strong reasoning supports this statement, which is deductive logic. At the same time, please note that if A is false, the above equation does not clarify the status on B; also, for B is true, it says nothing about A. These are the cases where additional information is required for weaker reasoning. The above statement also says that A implies B to mean that B is logically deducible from A. But in the formal logic, A implies B means only that the propositions A and AB have the same truth values. In general, whether B is logically deducible from A does not depend only on propositions A and B. It depends on the totality of propositions $(A, A', A'',\ldots)$ that we accept as true. Therefore merely knowing the propositions B and B alone or establishing that A and B both are true does not provide enough information to decide whether either is logically deducible from the other. Hence, the word implies can lead to a serious error, if not properly understood. From the two propositions A and B, others may be defined namely: (i) logical product (or conjunction): AB; (ii) logical sum (or disjunction): $A + B$; (iii) implication $A \Rightarrow B$; and (iv) negation $\bar{A}$. By combining these operations, repeatedly in every possible way, one can generate any new number of propositions, such as

$$C \equiv (A+\bar{B})(\bar{A}+A\bar{B})+AB(A+B) \tag{1.20}$$

The above combination leaves series of questions namely: (i) how large is the class of new propositions? (ii) Is it finite? (iii) Does it require connectivity? (iv) What is the smallest set of operations? All these questions are in general summarized as to whether the combinations of propositions increase the number of functions or decrease the number of operations? Let us say that $C = \text{fn}\ (A,B)$ and C can take values of either true or false; then the possible combinations are shown in the following table:

A	T	F
$f_1(A)$	T	T
$f_2(A)$	T	F
$f_3(A)$	F	T
$f_4(A)$	F	F

But one can easily see that the above combinations can be simplified as given below:

$$f_1(A) = A$$
$$f_2(A) = A + \bar{A}$$
$$f_3(A) = \bar{A}$$
$$f_4(A) = A\bar{A}$$

The above simple illustration derives two basic qualitative rules namely: (i) degree of plausibility is going to be represented by real numbers; and (ii) correspondence to common sense shall be established, which shall be in close agreement with the human thinking process.

Let us consider an example. Based on the prior information "C," evaluate A. If we add more information to C, then A becomes more plausible. This can be written mathematically as

$$A \mid C > A \mid C' \tag{1.21}$$

$$B \mid AC = B \mid AC';\ \ B \text{ does not change} \tag{1.22}$$

Then,

$$AB \mid C' > AB \mid C \tag{1.23}$$

Thus, different ways of reasoning yield the same result and hence it is always better to use all information to arrive at the result or conclusion. Two quantitative rules are very useful in plausible reasoning namely the product rule and the sum rule.

$AB \mid C$ means that both A and B are true for given C. The product rule is as follows:

$$W(AB \mid C) = W(B \mid C) * W(A \mid BC) \tag{1.24}$$

where W is a monotonic, continuous function for which derivatives exist.

For A is certain (true), one can expect the following:

$$A \mid C = A \mid BC \tag{1.25}$$

$$AB \mid C = B \mid C \tag{1.26}$$

Then,

$$W(AB \mid C) = W(B \mid C) * W(A \mid BC) = W(B \mid C) \tag{1.27}$$

Also,

$$W(A \mid BC) = 1 = W(A \mid C) \tag{1.28}$$

The sum rule is as follows:

$$A \mid C = F(\tilde{A} \mid C) \tag{1.29}$$

$$W^m(A) + W^m(\tilde{A}) = 1 \tag{1.30}$$

$$W^m(AB \mid C) + W^m(A \mid BC) * W^m(B \mid C) = 1 \tag{1.31}$$

Let W^m be called by probability, then we can rewrite as follows:

$$P(AB \mid C) = W^m(A \mid BC) * W^m(B \mid C) \tag{1.32}$$

$$P(A \mid C) + P(\tilde{A} \mid C) = 1 \tag{1.33}$$

$$P(A + B \mid C) = 1 - P((\tilde{A}\tilde{B}) \mid C) = P(A \mid C) + P(B \mid C) - P(AB \mid C) \tag{1.34}$$

If A and B are mutually exclusive, then $P(AB \mid C) = 0$. If A and B are mutual exclusive and exhaustive, then the following relationship holds good:

$$P\sum A_i \mid C = \sum P(A_i \mid C) = 1 \tag{1.35}$$

The above equation leads to the *Total probability theorem*. Also,

$$P(A_i \mid C) = \frac{1}{N} \tag{1.36}$$

Please note that the above equation does not decide the shape of the function, P. The above expressions can be further simplified. $P(A \mid C)$ is used when proposition A and B are used; or $p(A \mid C)$ is used when X and Y are real number, instead of propositions. Thus, probability is a function of (plausibility of) proposition A, which has a finite set of propositions.

Let us now examine the policeman example, again. For the logical reasoning,

$$A \Rightarrow B$$

$$\frac{A \text{ is True}}{B \text{ is True}}$$

$$\text{Result is } P(B \mid AC) = 1 \tag{1.37}$$

$$\text{Also } P(B \mid AC) > P(B \mid C) \tag{1.38}$$

The additional information, C has improved knowledge status of A, making it more plausible. C can be seen as an example as experience of the policemen. This leads to the conclusions that

$$P(A \mid BC) > P(A \mid C) \tag{1.39}$$

1.9 DEDUCTIVE REASONING

Based on the above discussions, the following statements are valid:

if A is true, then B is true. It is also interesting (and important) to note that A is true and therefore B is true. However, such deductive logic cannot be readily applied to examine reliability of the offshore structures due to the higher order of uncertainties; in reality, such analysis deals (or is forced to deal with) weaker reasoning. The following is the deductive reasoning:

If A is true, then B is true; for B is true, A becomes more plausible. If B is false, then A becomes less plausible. Let us try to understand this with the following example:

A: it will start to rain by 10 am at the latest
B: the sky will become cloudy before 10 am

The above events (statements) show that if A is true, B expresses only as the logical consequence of A and not a causal physical consequence; this leads to a weaker reasoning. Alternatively, if A is true, then B becomes more plausible. If B is true, A becomes more plausible. Therefore, in plausible reasoning, the judgment is not only to decide whether something becomes more or less plausible but also to evaluate the degree of its plausibility in some way. Referring to the above example, plausibility of rain by 10 am strongly depends on the darkness of the clouds at 9.45. Hence, in reasoning one is very much dependent on prior information in order to decide the degree of plausibility. This reasoning process takes place in almost the subconscious state of judgment.

1.9.1 Quantitative Rules

1.9.1.1 Product Rule

We seek a consistent rule relating the plausibility of $AB \mid C$ to those of $A \mid C$ and $B \mid C$ separately. The process of deciding that AB is true can be broken down into two parts: decide that B is true $B \mid C$. Having accepted that B is true, decide that A is true $A \mid BC$. Or equivalently, decide that A is true $A \mid C$; having accepted that A is true decide that is true $B \mid AC$. Formally we can state this as follows:

$$AB \mid C = F[B \mid C, A \mid BC] \tag{1.40}$$

The rule of agreement implies that given any change in prior information, such that B becomes more plausible but A does not change. This can be expressed as

$$B \mid C'' > B \mid C \tag{1.41}$$

$$A \mid BC = A \mid BC'' \tag{1.42}$$

By simple observation, one can state that AB becomes only more plausible and not less plausible. Under the above conditions

$$A \mid BC'' = A \mid BC \tag{1.43}$$

Introducing the real number, which is given by $X = B \mid C$, $y = A \mid BC$, the function F can be written as $F(x, Y)$. This results in $F(X, Y)$ being a continuous and monotonic increasing function of X and Y. This function has to be continuous to prevent a large increase in the plausibility of AB due to a small increase of plausibility of $A \mid C$ or $B \mid C$. This implies the following:

$$F_1(x, y) = \frac{\partial F}{\partial x} \tag{1.44}$$

$$F_2(x, y) = \frac{\partial F}{\partial y} \tag{1.45}$$

where F_i denotes differentiation with respect to the ith argument of F.

For example, one is interested to know the plausibility (ABC-D) that the three statements are simultaneously true because of the fact that Boolean algebra is associative. This can be evaluated in two ways. One way is to consider BC as a single statement and is given by

$$ABC \mid D = F[BC \mid D, A \mid BCD] = F\{F[C \mid D, B \mid CD], A \mid BCD\} \tag{1.46}$$

Alternatively, AB is considered as a single statement and is expressed as

$$ABC \mid D = F[C \mid D, AB \mid CD] = F\{C \mid D, F[A \mid BCD, B \mid CD]\} \tag{1.47}$$

In such case, the following statement holds good:

$$F[F(x,y),z] = F[x,F(z,y)] \tag{1.48}$$

The above equation is known also as the *associativity equation*. It is evident that the above equation has a trivial solution, that is $F(x, y)$ is constant. However, as this solution violates the monotonic requirement, it is of no use. By using the following abbreviations, we get

$$u \equiv F(x,y) \tag{1.49}$$

$$v \equiv F(y,z) \tag{1.50}$$

Therefore, Equation 1.48 will be reduced to the following form:

$$F(u,z) = F(x,v) \tag{1.51}$$

Differentiating the above equation with respect to x and y, we obtain as follows:

$$F_1(u,z)F_1(x,y) = F_1x,v)F_1(u,z)F_2(x,y) = F_2(x,v)F_2(y,z) \tag{1.52}$$

This leads to the following statement:

$$\frac{F_2x,v)F_1(u,z)}{F_1(x,y)} = \frac{F_1x,v)}{F_2(x,y)} \tag{1.53}$$

Defining the notation,

$$G(x,y) \equiv \frac{F_2(x,y)}{F_1(x,y)} \tag{1.54}$$

we can write the above equation as follows:

$$u = G(x,v)F_1(y,z) = G(x,y) \tag{1.55}$$

The above equation can be rewritten as

$$V = G(x,v)F_2(y,z) = G(y,Z)G(x,y) \tag{1.56}$$

Denoting the left-hand sides of above equations by U and V, respectively, one can write as follows:

$$\frac{\partial u}{\partial z} = \frac{\partial G(x,v)F_1(y,z)}{\partial z} = \frac{\partial G(x,y)}{\partial z} = 0 \tag{1.57}$$

$$\frac{\partial u}{\partial z} = \frac{\partial G(x,v)\partial F(y,z)}{\partial y \partial z} = \frac{\partial G(x,v)F_2(y,z)}{\partial y} = \frac{\partial v}{\partial y} \tag{1.58}$$

This implies that $V = G(x, y)G(y, z)$ is independent of Y. The most general function $G(x, y)$ with this property is given by

$$G(x,y) = r\frac{H(v)}{H(y)} \tag{1.59}$$

where r is a constant and $H(x)$ is arbitrary. Since F is a monotonic function, for $G > 0$ it is required that also $R > 0$. Based on the above equations, one can arrive at the following:

$$F_1(y,z) = \frac{H(v)}{H(y)} \quad F_2(y,z) = r\frac{H(v)}{H(z)} \tag{1.60}$$

Therefore, the relation $dv = dF(y,Z) = F_1dy + f_2dz$ takes the form:

$$\frac{dv}{H(v)} = \frac{dy}{H(y)} = \frac{dz}{H(z)} \tag{1.61}$$

It can be shown that a nontrivial solution for the above equation is in the following form:

$$w[F(x,y)] = w(x)w(y) \tag{1.62}$$

By introducing $x = B \mid C$ and $y = A \mid BC$, the above equation can be rewritten as

$$p(AB \mid C) = p(B \mid C)p(A \mid BC) \tag{1.63}$$

Equation 1.63 is termed as the *product rule*. By its construction, it is seen that $P(\cdot)$ should be a positive, continuous, monotonic function, which can be either increasing or decreasing. Now let us consider the limiting cases. The first one is the case in which $A \mid C$ is certain, satisfying the following condition:

$$AB \mid C = B \mid C \text{ and } \mathrm{A} \mid BC = A \mid C \text{ are true} \tag{1.64}$$

By expanding the above relationship using the product rule, we get

$$P(A \mid BC)p(B \mid C) = p(A \mid C)p(B \mid C) = p(B \mid C) \tag{1.65}$$

The above equation results in $p(A \mid C) = 1$, which is certainty. Alternatively, a case corresponds to $A \mid C$ is impossible in which the following condition is necessary:

$$AB \mid C = B \mid C \text{ and also } A \mid BC = A \mid C \tag{1.66}$$

By expanding the above using the product rule, we get

$$p(A \mid C)p(B \mid C) = P(B \mid C) \tag{1.67}$$

This holds for two values of $p(A \mid C)$, 0 and $+\alpha$. If we choose the solution $p(A \mid C) = 0$ as a convention, this results in $0 \leq p(x) \leq 1$

1.9.1.2 Sum Rule

Let the plausibility of A' be related to the plausibility of A. It can be easily shown that the functional form is given by

$$p(A \mid B) + p(A' \mid B) = 1 \tag{1.68}$$

It is interesting to assess whether these set of rules are adequate to decide the plausibility of any logic function $f(A_1, A_{2,\ldots,} A_n)$ of propositions $\{f(A_1, A_{2,\ldots,} A_n\}$. Let us seek a general formula for the logical sum $A + B$ by applying repeatedly the product and sum rules as explained below:

$$\begin{aligned} p(A+B \mid C) &= 1 - p(A'B' \mid C) \\ &= 1 - p(A' \mid C)p(B' \mid A'C) \\ &= 1 - p(A' \mid C)[1 - p(B' \mid A'C)] \\ &= 1 - p(A' \mid C) + p(A' \mid C)p(B' \mid A'C) \\ &= p(A \mid C) + p(A' \mid BC)p(B \mid C) \\ &= p(A \mid C) + [1 - p(A \mid BC)]p(B \mid C) \\ &= p(A \mid C) + p(B \mid C) - p(B \mid C)p(A \mid BC) \\ &= p(A \mid C) + p(B \mid C) - P(AB \mid C) \end{aligned} \tag{1.69}$$

The generalized sum rule is one of the most useful relationships.

1.9.2 Qualitative Rules

Now let us examine the relationship of the above theory to the deductive logic and different forms of plausible reasoning, which is presented in the earlier section. The deductive logic is based on the fact that if A is true, then B is also true. Then, following relationship holds good:

$$p(B \mid AC) = p(AB \mid C) / p(A \mid C) = p(A \mid C) = 1 \tag{1.70}$$

If A is true, then B is true $\equiv A \Rightarrow B$. If B is false, then A is false. Then, we get

$$p(A \mid BC) = \left[p(BA|C)/P(B|C)\right] = 0 \tag{1.71}$$

If A is true, then B is true $C \equiv A \Rightarrow B$. Also, if B is true, A becomes more plausible. Hence, we get

$$p(A \mid BC) = p(A \mid C)[P(B \mid AC) / p(B \mid C)] \tag{1.72}$$

But since $p(B \mid AC) = 1$ and $p(B \mid C) \leq 1$, we get

$$p(A \mid BC) \geq p(A \mid C) \tag{1.73}$$

Also, following statements hold good:

$$p(B \mid AC) \geq p(B \mid C) \tag{1.74}$$

$$p(A \mid BC) = p(A \mid C)[p(B \mid AC / p(B \mid C)] \tag{1.75}$$

$$P(A \mid BC) \geq p(A \mid C) \tag{1.76}$$

In the above sections, we discussed how plausible reasoning is connected to probability. Now we shall appreciate that reliability is "to assess in the mathematical way, the confidence one has in the occurrence of an event (of failure)." Let us understand this statement with the following examples:

EXAMPLE 1.1

Let the event E (given) be "It will rain tomorrow." $P[E]$ is the probability that it will rain tomorrow. $P[E]$ is a simple number to express the confidence that one has in the occurrence of the event. It translates the confidence in a mathematical way. Probability translates mathematically and consistently (no mistakes are made) the confidence into occurrence of any event of interest, E. This is called subjective probability. The question for self-reasoning is that what is the difference between plausibility and probability?

EXAMPLE 1.2

The knowledge status changes the confidence on occurrence of the event. Say, for example, the Google weather report is published and one is aware of it and let this be H, then,

$$P[E] = P(E \mid H)$$

Thus the knowledge status changes the probability of the event E. This reinforces the confidence level one has in the event E. However, it is always conditional on some hypothesis.

In estimating the event E, there exists a series of alternative events; these are called the space of events. With reference to Example 1.1, the space of events shall be considered as {cloudy, C; sunshine, S; fog, F}. Now $P(S) = 1$, where is the union of all possible events, related to the event under discussion, say rainfall, E. Thus,

$$S = \mathrm{U}\{EUCUSUF\}$$

In the considered example, please note that all the events are mutually exclusive. Therefore

$$P[EUA] = P[A] + P[E]$$

where the space of all other events is indicated by S. If they are not mutually exclusive, then the above statement is not true. In assessing the reliability of offshore structures, the space of events identified, say, for example, can be the compressive strength of concrete, tensile strength of steel, etc. They are not mutually exclusive because the capacity of reinforced concrete platforms depends on both the identified space of events.

To make the statement clear *knowledge status improves probability*, let us consider an example. Let us assume that there are two events, E and A; E is statistically dependent on A, implies that knowledge on A changes status on E. In such cases, the following statement is valid:

$$P(E \mid A) \neq P(E)$$

Let E be the event of rain tomorrow and $P[E]$ be 0.5; knowledge on the weather report predicts that tomorrow is cloudy. Now, $P[E] = 0.7$. Suppose, on the following day, in the morning, if the clouds become dark, then $P[E] = 0.8$. This can be expressed as follows:

$$P(E \mid C) \neq P[E \mid \text{google}] \neq P[E]$$

$P(E \mid C)$ is called *conditional probability*. Kindly note that C is improving knowledge status on E and therefore $P[C] = 1$. In such cases, the following equation is also valid:

$$P(E \mid C) = \frac{P[E \cap C]}{P[C]}$$

$$P[E \cap C] = P[E \mid C] * P[C] = P[C \mid E] * P[E] \quad \{\text{Product rule}\}$$

$$P[E \cup C] = P[E] + P[C] - P[E \cap C]$$

Total probability theorem states that

$$P[E] = P[E \mid C] * P[C] + P[E \mid \bar{C}] * P[\bar{C}]$$

1.10 CONTINUOUS PROBABILITY DISTRIBUTION FUNCTIONS

While the above discussions are limited to a discreet set of propositions, offshore structures deal with variables that are continuous in space and time. Let f is a continuous variable of interest, then the propositions are as follows:

$$F' \equiv (f \leq q), F'' \equiv (f > q) \tag{1.77}$$

As they are discrete, mutually exclusive and exhaustive, all rules derived earlier are applicable. Given some information on Y, probability of F' will depend on q, defining a function as given below:

$$G(q) \equiv P(F' \mid Y) \tag{1.78}$$

The above shall monotonically increase as a function of q. One will now be interested to know the probability that f lies within a specified interval. To understand this, let us define the propositions as

$$A \equiv (f \le a), B \equiv (f \le b), W \equiv (a \le f \le b) \tag{1.79}$$

Further, using Boolean algebra one can state that $B = A + W$. Since A and W are mutually exclusive, the sum rule reduces to the following:

$$P(B \mid Y) = P(A \mid Y) + P(W \mid Y) \tag{1.80}$$

But by definition, $P(B \mid Y) = G(b)$ and $P(A \mid Y) = G(a)$. Hence, the following statement is valid:

$$P(a \le f \le b \mid Y) = P(W \mid Y) = G(b) + G(a) \tag{1.81}$$

As $G(q)$ is continuous and differentiable, one can express it as follows:

$$P(a \le f \le b \mid Y) = \int_a^b g(f)df \tag{1.82}$$

where $g(f) = G''(f) \ge 0$ is the derivative of G, which is generally called the probability distribution function or the probability density function for f, given Y. The integral $G(f)$ is called the cumulative distribution function for f. Interestingly, the above equation leads to ambiguity when $G(q)$ is discontinuous at point q_0. Hence, it is important to note that the function approaches the limit in such a way that the density function develops a sharper and sharper peak, going in the limit of delta function $\{p_0\delta(q-q_0)\}$. This signifies a discreet hypothesis (H_o) while the enclosing (limiting) area is equal to the probability (p_0) of that hypothesis.

1.11 TESTING OF HYPOTHESES

Let A represent the proposition that "the fraction of bad gadgets is in the range $(f, f + df)$," then prior probability density function (pdf) can be calculated as

$$P(A \mid X) = d(f \mid X)df \tag{1.83}$$

which gives the probability that the fraction of bad gadgets is in the range $(f, f + df)$. Let D represent the result of the experiments conducted to identify the good (G) and bad gadgets (B). Then the posterior pdf for f can be calculated as

$$P(A \mid DX) = P(A \mid X)\frac{P(D \mid AX)}{P(D \mid X)} = g(F \mid DX)df \tag{1.84}$$

Therefore,

$$g(f \mid DX) = g(A \mid X)\frac{P(D \mid Ax)}{P(D \mid X)} \tag{1.85}$$

In the above equation, the denominator is a normalizing constant, which can be computed directly; but it is usually easier to determine it from the condition:

$$P(0 \le f \le 1 \mid DX) = \int_0^1 g(f \mid DX)df = 1 \tag{1.86}$$

Kindly note the evidence of the data lies in $P(D \mid AX)$. Since f is the only varying parameter and $P(D \mid AX)$ is a continuous function of f, $P(D \mid AX)$ can be replaced with $P(D \mid H_f X)$. At any random pick, if f is the fraction of bad gadgets, then the probability for picking a bad one at each trial would be f while that of getting a good one would be $1–f$. Probability at each trial are by hypothesis (hidden in X), which are logically independent. Given f, following statement holds good:

$$p(D \mid H_f X) = f^n (1-f)^{N-n} \tag{1.87}$$

It is important to note that the factors do not appear as we have knowledge of the order in which the results appear. The posterior pdf can now be calculated as

$$p(D \mid H_f X) = \frac{f^n (1-f)^{N-n} g(f \mid x)}{\int_0^1 f^n (1-f)^{N-n} g(f \mid x)df} \tag{1.88}$$

For example, in a multiple hypotheses test, prior pdf is given by

$$g(f \mid x) = \frac{10}{11}(-10^{-6})\delta\left(f - \frac{1}{6}\right) + \frac{1}{11}(1-10^{-6})\delta\left(f - \frac{1}{3}\right) + 10^{-6}\delta\left(f - \frac{99}{100}\right) \tag{1.89}$$

This is a case where the cumulative function of $\{G(F)\}$ is discontinuous while the three delta function corresponds to three hypotheses B, A, and C, respectively. Suppose that there is no prior knowledge except for the assurance that it possible for a machine to make a good or a bad gadget, one will be interested to know what prior probability should be assigned. One way is to assume that $g(f \mid X)$ is a constant. In order to normalize it correctly, one must consider the following:

$$g(f \mid x) = 1, \quad 0 \le f \le 1 \tag{1.90}$$

Equation 1.87 then reduces to a function known as complete beta function and is given by

$$p(D \mid H_f X) = \frac{(N+)!}{n!(N-n)!} f^n (1-f)^{N-n} \tag{1.91}$$

The right-hand side of the above equation has a peak within $0 \le f \le 1$, located by differentiation at $f = f' \equiv n/N$. To determine the sharpness of the peak, one can expand $L(f)$ in a power series about f' as given below:

$$L(f) = L(f') - \frac{(f-f')^2}{2\sigma^2} + \cdots \tag{1.92}$$

$$\sigma^2 \equiv \frac{f'\,(1-f')}{N} \tag{1.93}$$

and therefore, the following relationship holds good:

$$g(f \mid DX) = \exp L(f) \cong K \exp\left\{\frac{(f-f')^2}{2\sigma^2}\right\} \tag{1.94}$$

It is seen that the above follows a standard Gaussian or normal distribution.

1.12 SIMPLE AND COMPOUND HYPOTHESES

The hypotheses (A, B, C, H_f) that are discussed above refer to a single parameter ($f = (M/N)$). Such hypotheses are called simple hypothesis because in the space Ω of all parameters that are possible, f is represented by a single point. However, in some cases, one will be interested to test the hypothesis for the parameters lying within some subset of the space Ω while a particular value for f is uninteresting (Cartwright and Longuet-Higgins, 1956). For example, if $f > 0.1$, then one needs to take some action (e.g., machine may be allowed running) otherwise within the space $\Omega = [0,1]$ While the subset of interest is $\Omega = [0.1,1]$, the actual value of f is not of interest; this is called as nuisance parameter. As the chosen problem has no other parameter other than the variable (f) and their different intervals (df) are mutually exclusive, a discreet sum rule is applicable:

$$P(A_1 + A_2 + \cdots + A_n \mid DX) = \sum_i P(A_i \mid DX) \tag{1.95}$$

The above can be generalized into an integral as the variable (A_i) becomes more numerous. In that case, one can get rid of the nuisance parameter by integrating it as given below:

$$P(\Omega_1 \mid DX) = \frac{\int_{\Omega_1} f^n (1-f)^{N-n} g(f \mid x) df}{\int_{\Omega} f^n (1-f)^{N-n} g(f \mid x) df} \tag{1.96}$$

In case a uniform prior pdf for the single parameter (f) exists, then the above integral reduces to the incomplete beta function as given below:

$$P(a < f < b \mid DX) = \frac{(N+1)!}{n!(N-n)} \int_a^b f^n (1-f)^{N-n} df \tag{1.97}$$

when the hypotheses become numerous, one needs to follow a different approach. A set of discrete hypotheses can always be identified by one or more numerical indices that can identify them. In such cases, hypothesis testing transforms into the problem of estimation.

1.13 URN DISTRIBUTION

Let us now consider a problem statement of urn distributions. Suppose there are N number of balls in the urn that has both red and white color balls. Let r be red balls; number of white balls will be ($N–r$), which are unknown. In order to find the number of R balls (or white balls), one has to make a "predate" inference about their proportion are likely to be drawn in n draws from the urn. Interestingly, one can also invert this as a post-data problem. The data $D \equiv (n,r)$ are known but the contents (N,r) of the urn are not. Recalling the basics of sampling distribution, the following statement is valid:

$$p(D \mid NRI) = h(r \mid N, R, n) = \frac{\binom{M}{r}\binom{N-M}{n-r}}{\binom{N}{n}} \tag{1.98}$$

where I denotes the prior (known or assumed) information. If we succeed in n drawing N balls from the urn, then we can deduce that $N \geq n$. Using Bayes theorem, we get

$$p(N \mid DI) = p(N \mid I) \frac{p(D \mid NI)}{p(D \mid I)} \tag{1.99}$$

One shall accept that for every $p(N \mid I)$ that has been assigned, data would truncate the impossible values for N, leaving the relative probabilities of the possible values unchanged. In that case, we get

$$p(N \mid DI) = \begin{Bmatrix} Ap(N \mid I) & \text{if} & N \geq n \\ 0 & \text{if} & 0 \leq N \leq n \end{Bmatrix} \tag{1.100}$$

One shall also conclude as follows:

$$D \equiv N \geq n$$

Then

$$p(D \mid NI) = \begin{Bmatrix} 1 & \text{if} & N \geq n \\ 0 & \text{if} & 0 \leq N \leq N \end{Bmatrix} \tag{1.101}$$

By applying Bayes' theorem, we get

$$p(D \mid NI) = p(N \mid I)\frac{p(D \mid NI)}{p(D \mid I)} = \begin{Bmatrix} Ap(N|I) & \text{if} & N \geq n \\ 0 & \text{if} & 0 \leq N \leq n \end{Bmatrix} \tag{1.102}$$

where $A = p(D \mid I)^{-1}$ is the normalizing constant. However this will not be the case unless $p(D \mid NI)$ is independent of N for $N \geq n$. The general condition that the data tells nothing about N except truncating the impossible values will lead to a nontrivial condition on $p(R \mid NI)$, as given below:

$$p(D \mid NI) = \sum_{R=0}^{N} p(D \mid NRI)p(R \mid NI) = \begin{Bmatrix} f(n,r) & \text{if} & N \geq n \\ 0 & \text{if} & 0 \leq N \leq n \end{Bmatrix} \tag{1.103}$$

where $f(n, r)$ may depend on the data but is independent of N. For the case of hypergeometric distribution, the following holds good:

$$\sum_{R=0}^{N} \binom{R}{r}\binom{N-R}{n-r} p(R|NI) = F(n,r)\binom{N}{n} N \geq n \tag{1.104}$$

The posterior probability distribution for R can be written as

$$p(R \mid DNI) = p(R \mid NI)\frac{p(D \mid NRI)}{p(D \mid NI)} \tag{1.105}$$

It is also interesting to note that different choices for the probability $p(R \mid DNI)$ will lead to different results. Considering a case where the knowledge status on the prior information of red balls is shown by I_0, then $p(R \mid DNI_0)$ is derived as follows:

$$p(R \mid DNI_0) = p(R \mid NI_0)\frac{p(D \mid NRI_0)}{P(D \mid NI_0)} \tag{1.106}$$

Then $p(R \mid DNI_0)$ can be calculated as below:

$$p(R \mid DNI_0) = \frac{1}{N+1}\frac{\binom{R}{r}\binom{N-R}{n-r}}{\binom{N}{n}p(D|NI_0)} \tag{1.107}$$

where $p(D \mid NI_0)$ can be calculated as

$$p(D \mid NI_0) = \sum_{R=0}^{N} p(R \mid DNI_0)p(D \mid NI_0) = \sum_{R=0}^{N}\frac{\binom{R}{r}\binom{N-R}{n-r}}{\binom{N}{n}(N+1)} \tag{1.108}$$

It can be also shown that the following relationship holds good:

$$\sum_{R=0}^{N}\binom{R}{r}\binom{N-R}{n-r} = \binom{N+1}{n+1} \tag{1.109}$$

By making the above relationship more general, $p(D \mid NI_0)$ will be simplified as follows:

$$p(D \mid NI_0) = \frac{\binom{N+1}{n+1}}{\binom{N}{n}(N+1)} = \frac{1}{n+1} \tag{1.110}$$

Substituting $p(D \mid NI_0)$ in Equation 1.108, posterior distribution for R is derived as follows:

$$p(D \mid NI_0) = \frac{\binom{R}{r}\binom{N-R}{n-r}}{\binom{N}{n}((N+1)/(n+1))} = \frac{\binom{R}{r}\binom{N-R}{n-r}}{((N+1)/(n+1))} \tag{1.111}$$

It is interesting to note that this is not a hyper-geometric distribution since the variable is R and not r. The expected value for R can be calculated as below:

$$R = \sum_{0}^{N} R_p(R \mid DNI_0) \tag{1.112}$$

In order to calculate the summation in Equation 1.110, one can use the following equality:

$$(R+1)\binom{R}{r} = (r+1)\binom{R+1}{r+1} \tag{1.113}$$

$$R+1 = (r+1)\binom{R+1}{r+1}^{-1}\binom{R+2}{r+2} = \frac{(N+1)(r+1)}{(n+2)} \tag{1.114}$$

Once the expected value for R is computed, one can calculate also the expected value of the fraction F of the red balls remaining in the urn, as below:

$$F = \frac{R-r}{N-n} = \frac{r+1}{n+2} \tag{1.115}$$

This is quite close to the value r/n for large R and n.

1.14 RANDOM VARIABLES

Any variable is considered as random not because the value of the variable assumes any random number (Madsen, 1988; Nigam and Narayanan, 1994). The value assigned to the random variable is not random but fixed; randomness is due to the fact that the assigned value is not known. As it is being guessed in the wide range of possible values, these variables are termed as random variables (Papoulis and Pillai, 1991, 2002). A random number is just an association of a number to any specific event in a given space. A random number is therefore a representation of the event that has randomness associated with itself. It is only a convention to deal with the random event, mathematically.

1.14.1 Generation of Random Numbers with a Specified Distribution

A sequence of random numbers x_i, $i = 0,1,2,\ldots$ from a nonuniform probability distribution defined by the cumulative density function (cdf) $F_x(x)$ can be generated as follows. First, generate a sequence of uniform (0,1) random numbers, where u_i, $i = 0,1,2,\ldots$. Then compute as below:

$$X_i = F - x(u_i), \quad i = 0,1,2, \tag{1.116}$$

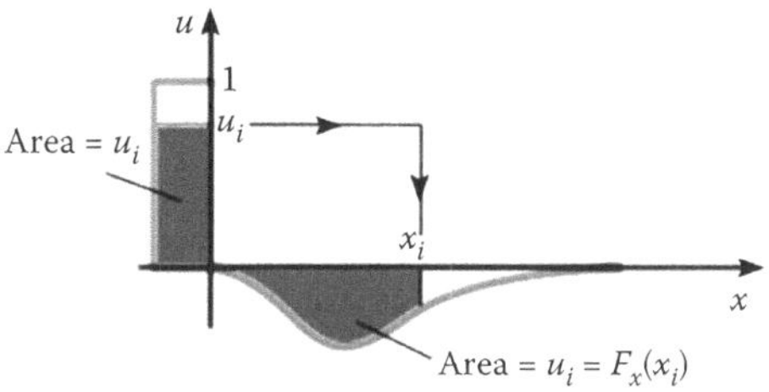

FIGURE 1.1 Plot of the variables.

This procedure operates by setting the area corresponding to $u \leq u_i$ under the uniform pdf, which is equal to the area corresponding to $x \leq x_i$ under the PDF of interest, as shown in Figure 1.1.

For X qualifying the standard normal distribution, a more efficient procedure is to generate pairs of uniformity, which is distributed in the range (0, 1) of random variables (u_1,u_2); the following equation is more relevant:

$$X_1 = \sqrt{-2\ln(u_1)}\ \sin 2\pi u_2 \tag{1.117}$$

$$X_2 = \sqrt{-2\ln(u_1)}\ \cos 2\pi u_2 \tag{1.118}$$

Please note that if the uniform random variables are statistically independent, then the standard normal random variables generated using the above equations will also be statistically independent.

1.14.2 Multiple Random Variables

To generate an outcome of $y = [y_1, y_2, y_3, \ldots\ y_n]T$, which is joint random variables with mean vector M and covariance matrix Σ,, one needs to generate a sequence of n random variables that are statistically independent using the following relationship:

$$X = [x_1, x_2, \ldots, x_n]^T$$

$$y = Lx + M \tag{1.119}$$

where the lower triangular matrix is found from Cholesky decomposition. The above procedure is repeated for the number of outcomes desired.

1.14.3 Nataf Random Variables

To generate an outcome of $y = [y1, y2, y3, \ldots, yn]T$, which is a vector of Nataf-distributed random variables with marginal $F_{yi}(yi)$, $i = 1,2,\ldots,n$ and a correlation coefficient matrix [R] of size ($n \times n$), one needs to generate a sequence of n random variables that are statistically independent using the following relationship:

$$Z = L_o x \tag{1.120}$$

where L_o is a lower triangular matrix that satisfies the following relationship:

$$L_0 L_0^T = R_0$$

where R_o is the modified correlation coefficient matrix for the Nataf random variables. The outcome $y = [y_1, y_{2,\ldots,} y_n]^T$ is then determined as follows:

$$\begin{Bmatrix} y_1 \\ y_2 \\ \cdot \\ \cdot \\ \cdot \\ y_n \end{Bmatrix} = \begin{Bmatrix} F_{y_1}^{-1}(\Phi(Z_1)) \\ F_{y_2}^{-1}(\Phi(Z_2)) \\ \cdot \\ \cdot \\ .F_{y_n}^{-1}(\Phi(Z_3)) \end{Bmatrix} \tag{1.121}$$

The above procedure is repeated for as many outcomes of Y as desired.

1.14.4 Random Variables Defined by Their Conditional Distributions

To generate an outcome of $X = [x_1, x_2, \ldots, x_n]^T$ which is a vector of random variables whose joint PDF, $f_x(x)$ is defined by their conditional distributions, one need to generate a sequence of n statistically independent, uniformly distributed random variables $\{u_i, i = 1,2\ldots n\}$ and then apply the Rosenblatt transformations, as given below:
$x_1 = F_{x_i}^{-1}(u_1)$

$$x_2 = F_{x2|x1}^{-1}(u_2 \mid x_1)$$

$$x_n = F_{x_n|x_{n-1}x_{n-2}}^{-1} \ldots\ldots\ldots x_1\left(u_n \mid x_{n-1}x_{n-2}\ldots x_1\right) \tag{1.122}$$

where Equation 1.122 gives the conditional CDF of x_1 given $X_{i-l} = x_{i-1}$, $X_{i-2} = x_{i-2}, \ldots$, $X_1 = x_1$. The above procedure is repeated for as many outcomes of X as desired.

1.15 MONTE CARLO SIMULATION METHOD

As discussed earlier, evaluation of the following is a major concern:

$$P_f = \int_x I(x) f(X) dx \tag{1.123}$$

where, the indicator function is given by

$$I(x) = \begin{cases} 1 & \text{if} \quad x \in \Omega \\ 0 & \quad \text{otherwise} \end{cases} \tag{1.124}$$

and, in general,

$$\Omega = \cup_m \cap_{i \in c_m} \{g_i(x) \le 0\} \tag{1.125}$$

Let $q_l = I(Xi)$ where Xi is the $_i$th random sampling of X. Please note that $p(q_i = 1) = p_f$

and $p(q_i = 0) = 1 - p_f$. Furthermore, q_i, i = 1,2..., are statistically independent. Consequently, $q1$, $q2$, is a Bernoulli sequence and can be written as

$$[q_i] = 1(p_f) + 0(1 - p_f) = p_f \tag{1.126}$$

$$\begin{aligned} \text{Var}(q_i) &= E[q_i^2] - E^2[q_i] \\ &= 1(p_f) + 0(1 - p_f) - p_f^2 \\ &= p_f(1 - p_f) \end{aligned} \tag{1.127}$$

where $q_i^2 = q_i$ for Bernoulli variables. Using the definition of the expectation operator, one can also express the probability of failure as the expected value of the indicator function. Mathematically, it can be expressed as

$$P_f = E[I(x)] \tag{1.128}$$

Therefore, an estimator of the probability of failure is given by

$$P_f' = E'[I(x)] = \frac{1}{N}\sum_{i=1}^{N} q_i = \bar{q} \tag{1.129}$$

where $\bar{q}$ is the sample mean obtained from N simulation of the random vector X. It should be evident that $\hat{p}$ is a random variable. The expected value of this estimator is given by

$$\boldsymbol{E}\left[p_f'\right] = E\left[\frac{1}{N}\sum_{i=1}^{N} q_i\right] = \frac{1}{N}\sum_{i=1}^{N}[q_i] = \frac{1}{n}\left(N_{p_f}\right) = p_f \tag{1.130}$$

Thus, $\hat{p}_f$ is an unbiased estimator of P_f. The variance of $\hat{p}_f$ is given by

$$\text{Var}(P_f') = \text{Var}\left(\frac{1}{N}\sum_{i=1}^{N} q_i\right) = \frac{1}{N_2}\text{Var}\left(\sum_{i=1}^{N} q_i\right) = \frac{1}{N_2}\text{Var}(q_i) = \frac{1}{N}p_f(1-p_f) \tag{1.131}$$

The coefficient of variation of $\hat{p}_f$, which quantifies the accuracy of estimate, is given by

$$\delta = \sqrt{\frac{p_{f(1-p_f)/N}}{p_f}} = \sqrt{\frac{1-p_f}{Np_f}} \tag{1.132}$$

It is to be noted that δ decreases with the increase in the number of simulations. This implies that the estimate of the probability of failure improves as the analysis proceeds. The coefficient of variance can be used to decide when the simulation should be stopped. Typically, a tolerable or target coefficient of variance (COV), δ_{target}, is specified and the simulations are terminated when $\delta \leq \delta$ target. Usually, $0.01 \leq \delta_{\text{target}} \leq 0.05$ is taken for a typical simulation study. As an example, for a simulation in which $p_f = 0.01$ and $\delta_{\text{target}} = 0.05$, approximately 39,600 simulations are necessary. As the probability of failure is unknown prior to the analysis, the required number of simulations is also unknown until the analysis is complete. The accuracy of Monte Carlo simulations can be improved by two ways namely: (i) by increasing the number of simulations, *N*; and (ii) by increasing the probability of failure, P_f. On the best engineering judgment, one prefers the former as the latter will lead to a physical change in the problem under consideration.

1.16 IMPORTANCE OF SAMPLING

As reliability studies are probabilistic based, it is imperative to emphasize the importance sampling (Rajashekar and Ellingwood 1993). One of the methods of reducing the variance in the Monte Carlo estimate of the probability of failure of a component or system is by sampling. This involves in rewriting the probability of failure as given below:

$$P_f = \int_Z I(X)\frac{f(x)}{h(x)}h(x)dx \tag{1.133}$$

here $h(x)$ is the sampling density function and $I(x)\,f(x)/h(x)$ acts as the indicator function. It is necessary to note that the sampling density function is chosen to remain nonzero wherever $I(x)\,f(x) \neq 0$. This is due to the fact that no regions of the failure domain are excluded from the analysis. Using the definition of the expectation operator, one can rewrite as

$$p_f = E\left[I(X)\frac{f(x)}{h(x)}\right] \tag{1.134}$$

The vital point is to choose $h(x)$ such that sampling is done more frequently from the failure domain. Thus, the ideal sampling function is given by

$$h(x) = I(X)\frac{f(x)}{P_f} \tag{1.135}$$

For this choice, mean of the estimate is identical to p_f with a zero variance for any N. But this ideal situation is not practical as the probability of failure is what is computed from the whole exercise (Rosenblueth, 1975). Therefore, selecting the sampling function is a critical step in importance sampling. In fact, a poor choice of $h(x)$ can increase the variance of $\hat{p}_f$, thereby making a crude Monte Carlo simulation. Several sampling functions that have been proposed for importance sampling are described further. One of the methods is to propose sampling within a hyper-rectangle, which is centered about the design point (Shinozuka, 1983). The plot of the same is shown in Figure 1.2.

Unfortunately, the estimate of the failure probability obtained from this approach is biased as the procedure does not assign any sampling density to regions of failure domain.

$$\lim_{n\to\alpha}(E[p'_f] \neq P_f) \tag{1.136}$$

If it is known that the safe set in the standard normal space is concave, as shown in Figure 1.2, then sampling over the half-space is defined by

$$\alpha^T u \geq \beta \tag{1.137}$$

where α is the unit normal to the limit state surface at the design point and β is the first-order reliability index. To improve the accuracy of failure estimate, it is therefore necessary to improve the efficiency of the simulation method. This approach is shown in Figure 1.3.

The sampling function in this case is given by

$$h(u) = \begin{Bmatrix} \dfrac{\Phi_n(u)}{\Phi(-\beta)} \\ 0 \end{Bmatrix} \text{if } \beta - \alpha^T u \leq 0 \tag{1.138}$$

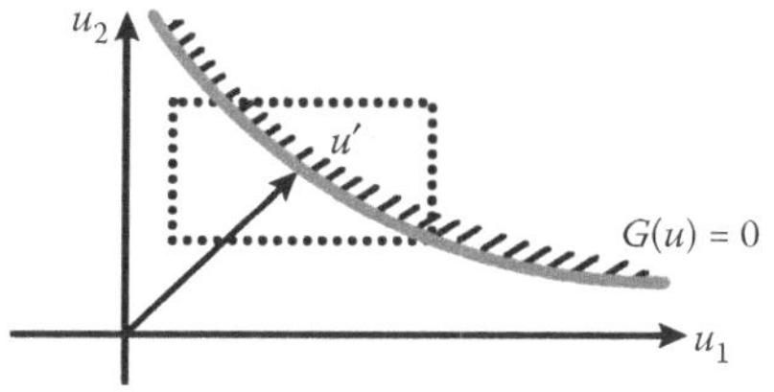

FIGURE 1.2 Sampling within hyper-rectangle.

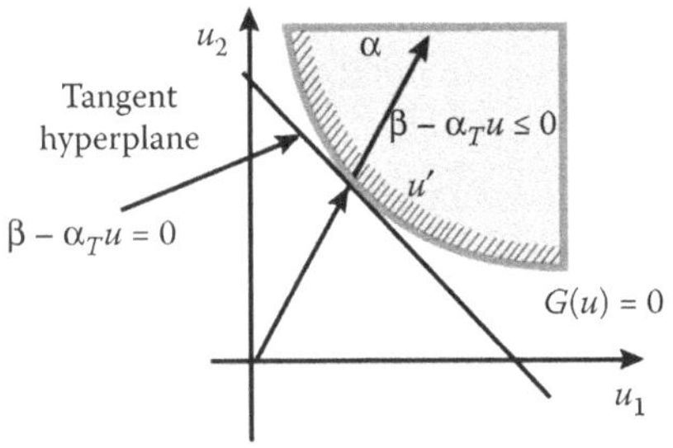

FIGURE 1.3 Improvement of failure estimate in two-dimensional domain.

In this approach, it is necessary to simulate $\{u\}$ such that the following condition is satisfied:

$$u \mid \beta - \alpha^T u \leq 0 \tag{1.139}$$

which can be done as follows:

Step 1. Simulate the standard normal variable u_i, $i = 1,2,3,\ldots\ (n-1)$

Step 2. Simulate a clipped normal random variable z from the PDF

$$f_z(z) = \frac{2}{\sqrt{2\pi}} \exp\left(-\frac{z^2}{2}\right),\ z \leq 0$$

Step 3. Solve for u_n from the following relationship:

$$Z = \beta - \alpha^T u$$

i.e.,

$$u_n = \frac{\beta - z - \sum_{i=1}^{n-1} \alpha_1 u_1}{\alpha_n}$$

Note that if the failure domain does not lie entirely in the half-space defined by $B - \alpha^T u \leq 0$, then this approach yields a biased estimate for the probability of failure. Figure 1.4 shows a typical case of failure domain.

If $\| u_i \| < \beta$ go to step 1

Else $q_i = I(u_i)[1 - x_m^2(\beta^2)],\ i = i + 1$

Go to step 1

Note that the counter i is incremented only when $| \mathbf{u_I} \| \geq \beta$.

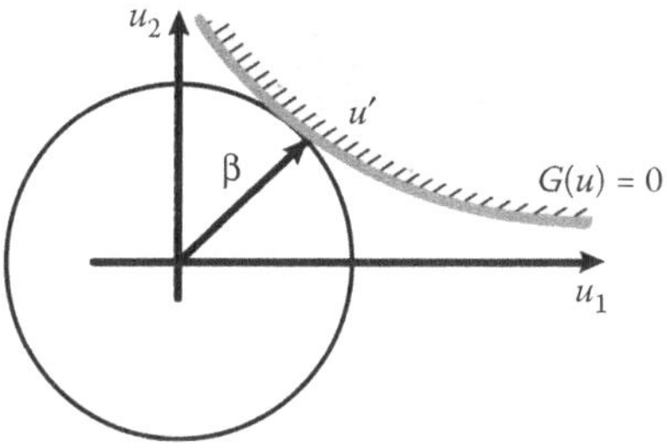

FIGURE 1.4 Typical case of failure domain.

At $i = N$, we get

$$p'_f = \frac{1}{N}\sum_{j=1}^{N} q_j \tag{1.140}$$

$$\delta \approx \sqrt{\frac{1-p'_f}{N_{p'_f}}} \tag{1.141}$$

The simulation is terminated when $\sigma \leq \sigma$ target or the prescribed maximum number of simulations is reached. It is also adequate if the sampling is done with the normal density at each design point (u_i^*), which is identified as shown in Figure 1.5.

In this case, for (m) number of limit state surfaces, the sampling density function is given by

$$h(u) = \sum_{i=1}^{m} w_i h_i(u) \tag{1.142}$$

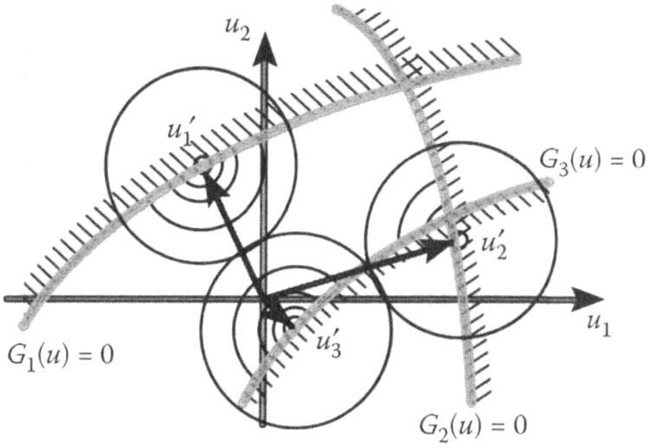

FIGURE 1.5 Sampling with normal density at design point.

$$h_i(u) = \frac{1}{(2\pi)^{n/2}\left(\det \sum i\right)^{1/2}} \exp\left[-\frac{1}{2}(u-u_i^*)^T \sum_i^{-1} (u-u_i^*)\right], -\alpha < u < \alpha \tag{1.143}$$

Equation 1.143 gives the multivariate normal PDF with mean vector u_i^* and covariance matrix $\Sigma_i = \sigma_i^2 I$ and w_i is a weight function ($\Sigma w_i = 1$) for the ith design point. Standard deviation (s_i, $i = 1,2,\ldots,m$) and weights (w_i, $i = 1,2,\ldots,m$) are chosen such that $h(u)$ approximates $f(x)$ in the failure region as closely as possible. It is quite usual that w_i will be larger for limit state functions with smaller reliability indices. One of the major advantages of this method is that it is unbiased even though it may not be as efficient as other approaches. This is due to the fact that in this method, approximately half of the sample points fall within the safe domain. However, if a failure region is not identified prior to the analyses, this method will indicate the unsuitability with a critical deviation of coefficient of variation from that of the standard value.

1.17 DIRECTIONAL SIMULATION

Directional simulation is another class of simulation method that is based on the concept of conditioning (Ale, 2002). Suppose the n vector of the random variables can be partitioned in the form:

$$X = [YT, ZT]^T$$

then, the following relationship holds true:

$$\begin{aligned} P_f &= P \cup_m \cap_{i=c_m} \{g_f(y, Z) \le 0\} \\ &= \int_x P(\cup_m \cap_{i=c_m} \{g_f(y, Z) \le 0\} Z) f_Z(Z) dz \end{aligned} \tag{1.144}$$

A special case of conditional simulation is directional simulation. In this case, the simulation is performed in the standard normal space with the vector of standard normal variate as given below:

$$U = Ra \tag{1.145}$$

where $R = \|u\|$ and $a = u/\|u\|$ is the unit direction vector. Thus, by conditioning on a, we can write as follows:

$$\begin{aligned} P_f &= P(\cup_m \cup_{i=c_m} \{G_i(u) \le 0\}) \\ &= P(\cup_m \cap_{i=c_m} \{G_i(Ra) \le 0\}) \\ &= \int_x P(\cup_m \cap_{i=c_m} \{G_i(Ra) \le 0\} \mid a) f_A(a) da \end{aligned} \tag{1.146}$$

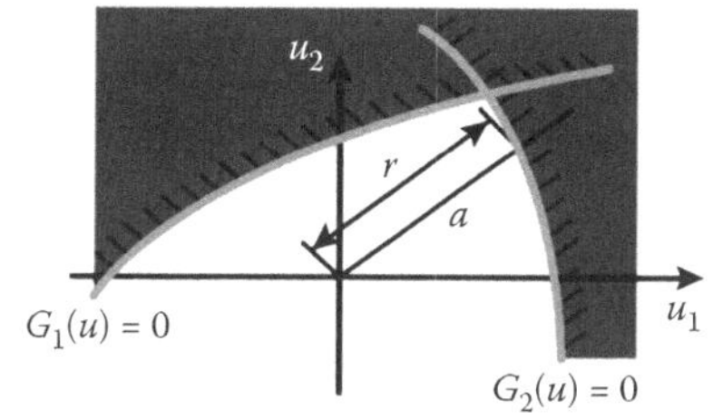

FIGURE 1.6 Failure boundary with CDF of chi-square distribution.

where $f_A(a)$ is the PDF of a, which is assumed to be uniform on the unit sphere, which is centered at the origin of the standard normal space. With the given fact that $R^2 = | u |^2$ has the chi-square distribution with n degrees of freedom, we get

$$P(\cup_m \cap_{i=c_m} \{G_i(Ra) \le 0\} \mid a) = 1 - x_m^2(r^2) \tag{1.147}$$

where $x_m^2(\cdot)$ is the CDF of the chi-square distribution with n degrees of freedom and r is the distance from the origin to the boundary of the failure domain along the direction (a), as shown in Figure 1.6.

For a set of simulations, $\{a_i, i = 1,2,\ldots,N\}$ of $f_A(a)$, an estimate of the probability of failure is given by

$$P_f' = \frac{1}{N}\sum_{j=1}^{1} q_{ji} \tag{1.148}$$

$$q_i = P(\cup_m \cap_{i=c_m} \{G_i(Ra) \le 0\}) \tag{1.149}$$

$$= 1 - X_n^2(r_j^2) \tag{1.150}$$

where r_j is the distance from the origin to the boundary of the failure domain in direction a_j. By employing directional simulation, probability of failure can be expressed as

$$P_{f=\int_a P(\cup_m \cap_{i=c_m} \{G_i(Ra)\le 0\})|a \frac{f_A(a)}{h(a)} h(a)da} \tag{1.151}$$

$$q_i = P(\cup_m \cap_{i=c_m} \{G_i(Ra) \le 0\}) \frac{f_A(a_j)}{h_{a_j}} \tag{1.152}$$

$$= [1 - x_m^2(r_j^2)] \frac{f_a(a_j)}{h(a_j)} \tag{1.153}$$

where $h(a)$ is the sampling density for a, which should be chosen such that the simulated directions are concentrated toward those regions of the sample space that provide the maximum information about the probability of failure (Bierager, 1990). A composite sampling density is given by

$$h^c(a) = w_1 h(a) + w_2 f_A(a) \tag{1.154}$$

where w_1 and w_2 are weights with $w_1 + w_2 = 1$. The purpose of this composite sampling density is to concentrate the simulations in the desired directions while retaining a small number of simulations uniformly distributed on the unit sphere. This is done to ensure that no failure domains are missed. As such, w_2 should be set to a small value relative to w_1 (e.g., $w_2 = 0.1$).

1.18 STATISTICAL THEORIES OF EXTREMES

In reliability estimates, parallel systems can also be estimated by the use of certain limit distributions known as extreme value distributions. One of the areas of application is to determine the strength of material under the worst combination of loads, maximum wind gusts, etc. (Box and Wilson, 1951; Bottelberghs, 2000). Let us consider an example to understand this problem. Let us have a system S, which consists of n components whose lifetime are $\{\{T_1 \ldots T_n\}$. These life time values are assumed to be independent and identically distributed with reliability function $R(t)$. Let T_{sys} be the lifetime of the system and is given by

$$T_{sys} = \min_{1 \le i \le n} T_i \tag{1.155}$$

Reliability of the system R_{sys} is therefore given by

$$\begin{aligned} R_{sys}(t) &= P(T_{sys} > t) \\ &= P(T_1 > t, T_2 > t, \ldots, T_n > t) \\ &= \Pi\ P(T_i > t) \\ &= R^n(t) \end{aligned} \tag{1.156}$$

By taking the limit $n \rightarrow \alpha$, it can be easily seen that if $R(t) < 1$, then $R_{sys}(t) = R^n(t) \rightarrow 0$, if $n \rightarrow \alpha$ In general, system reliability can be estimated within a given time scale by choosing two sequences $a_n > 0$ and b_n such that the sequence of stochastic variables is expressed as

$$\frac{T_{sys} - b_n}{a_n} \tag{1.157}$$

1.19 MODELING OF ENVIRONMENTAL LOADS

Loads that act on offshore structures (Srinivasan Chandrasekaran, 2015a, b) are classified as: (i) permanent or (dead) loads named as *P* class; (ii) operating or (live) loads named as *L* class; (iii) deformation loads named as *D* class; (iv) environmental loads including earthquake loads, named as *E* class; and (v) accidental loads, named as *A* class. Permanent loads include time-dependent loads. For example, gravity loads such as weight of the structure, weight of permanent installations, etc., fall under this category. While this class of loads can be estimated with higher accuracy, a characteristic value is used to address the associated uncertainties (Kinsman, 1965; Kiran, 2014, 2012). Live loads are associated with the nature and operational modes of the plants and equipment of the top side. Choice of the characteristic value depends on the type of the structure under consideration. Deformation loads are caused by the imposed deformation on the structure that arises from large displacement. These loads may also arise from deformations caused by temperature difference, differential settlement of foundation, settlement of supports that arise in case of Articulated Towers, Triceratops, etc. Characteristic values prescribed in the design should be the maximum possible value as this will be associated with a very high degree of uncertainty (Lin, 1967; Lin and Cai, 2004). Environmental loads arise from the ocean environment that include wave loads, wind loads, ice loads, etc. The mean return period is used to account for the characteristic value of this class of loads. Accident loads arise from exceptional conditions like collision, drop objects, explosion and fire, etc. The characteristic value of this class of loads depends on the subject of investigation, operational manner, etc. As stated above, a variety of uncertainties arise in all class of loads; characteristic values are used to account for these uncertainties (Maddox and Wildenstein, 1975). Factors that affect the uncertainties are namely: (i) variability within material quality with respect to strength and other engineering properties such as modulus of elasticity, area of cross section, etc.; (ii) variability in geometric properties of components that arise during the fabrication process; (iii) uncertainties in the specified loads with respect to change in sea state and operational conditions; (iv) idealization of real structure into a model for analysis; (v) accuracy of numerical or analytical computation; (vi) quality of workmanship; and (vii) loss of strength due to corrosion.

1.19.1 Return Period

Environmental loads are associated with one of the main uncertainties that arise from its reoccurrence with a similar magnitude within the service life of the structure. Offshore structures are very sensitive to this data as this is vital in assigning design loads to the structure (Mohd Umair, 2012; Nolte and Hansford, 1976). The mean return period is the expected number of years between which a given seasonal maximum is likely to occur. Say, for example, if the return period is 50 years, it means that the characteristic load is expected to occur only once in 50 years. Inverse of the return period is the probability that the extreme value of the event will be exceeded in any one year. For example, if the return period is 100 years (which is a common case in wind loads), probability that the characteristic value will be exceeded in any one year is 0.01 (=1/100). But as a designer, one may not be interested in the reoccurrence of any event with its extreme value. Rather, the interest is about the fact that what is the probability that the design load will be exceeded during the life of the structure? Since (1/*R*) is the probability that the characteristic load will be exceeded in 1 year (1–1/*R*) will be that it will not exceed in any one year. For the structure whose life is denoted as "*n*, probability that the extreme event will not be exceeded during those "*n* years" is the designer's interest; this is given as $(1-1/R)^n$. Hence, the probability of characteristics load to be exceeded, at least once in lifetime of the structure is given by

$$p_n = 1 - \left\{1 - \frac{1}{R}\right\}^n \tag{1.158}$$

For example, if the return period is 50 years of any characteristic load and the life time of the structure is 20 years (which is common in offshore structures as the oil reserve for exploration is not consistent beyond a maximum period of 20 years), then probability that the design load is exceeded at least once during the 20 years is given by

$$p_{20} = 1 - \left\{1 - \frac{1}{50}\right\}^{20} = 0.33 \tag{1.159}$$

It is clear from the above example that there is a chance of the design load exceeding its characteristic value at least by 33% at least once within the lifetime of the structure. As a designer, this is quite important as this percentage is significantly high for a reasonable estimate of the cited example. Therefore, safety factors are used in the design to take care of such uncertainties (Marshall, 1976; Pasman et al., 2009; Srinivasan Chandarsekaran, 2015c).

1.20 ESTIMATE OF DISTRIBUTION PARAMETERS

Probability distributions are a vital part of statistics. A few of the practical applications are: (i) to compute the confidence intervals for parameters and (ii) to calculate critical regions for hypothesis tests. For univariate data, it is often useful to determine a reasonable distributional model for the data. Statistical intervals and hypothesis tests are often based on specific assumptions of the type of distribution that follows. Before computing an interval or test based on this assumption, one needs to verify the justification of this assumption for the given data set. This does not mean that the distribution needs to be the best-fitting distribution for the given data, but adequate-enough models should be assumed so that the statistical technique yields valid conclusions. It is therefore important to use the simulation studies with random numbers that are generated using a specific probability distribution. There are various methods, both numerical and graphical, for estimating the parameters of a probability distribution; important ones are discussed below:

1.20.1 Method of Moments

Method of moments equates the sample moments to that of the parameter estimates. While the primary advantage is simplicity, the disadvantage is that they are often unknown; even known, they do not qualify to have the desirable properties of maximum likelihood and least squares estimators. The primary use of moment estimates is as only the initial approximations to estimate maximum likelihood and least squares estimates more precisely (Tvedt, 1990; Vendhan, 2004; Wirsching et al., 2006).

1.20.2 Maximum Likelihood

Maximum likelihood estimation begins with the likelihood function of the sample data. Likelihood of a set of data is the probability of obtaining that particular set of data from the given or chosen probability model. This contains unknown parameters whose values that maximize the sample likelihood are known as MLEs. Maximum likelihood provides a consistent approach to parameter estimation problems, which enables its use for a large variety of estimation situations. Further, maximum likelihood method has desirable mathematical and optimality properties. They show minimum unbiased variance estimates for large sample sizes. By considering a very large number of random samples to replace the original sample population, the average value will match closely to that of the population value. As this will show a minimum variance, it expresses the narrowest confidence interval of all estimators of that type. They follow normal distributions in general and hence sample variances can be used to generate confidence bounds and hypothesis tests for the parameters. Several popular statistical software packages provide excellent algorithms for MLEs for many of the commonly used distributions. This helps mitigate the computational complexity of maximum likelihood estimation. There are a few disadvantages as well. Likelihood equations are case-specific for a given distribution and estimation problem. The mathematics is often nontrivial, particularly if confidence intervals for the parameters are desired. Except for a few cases where the maximum likelihood formulas are in fact simple, it is generally best to rely on high-quality statistical software to obtain MLEs. Fortunately, high-quality maximum likelihood software is becoming increasingly common. MLEs can be heavily biased for small samples. They are also highly sensitive to the choice of starting values.

1.20.3 Least Squares

Nonlinear least squares provide an alternative to maximum likelihood. Nonlinear least squares software are available in many statistical software packages, which generally do not support MLEs. If the software provides nonlinear fitting and has the ability to specify probability function one is interested in, then one can generate least squares estimates for that distribution. This allows reasonable estimates for distributions even if the software does not provide MLEs. However, there are a few disadvantages. It is not readily applicable to a censored data and is also very sensitive to the choice of starting values.

1.20.4 Probability Plots

Probability plots can be used to estimate the shape parameter of a distribution with a single shape parameter. After determining the best value of shape parameter, a probability plot can be used to estimate the location and scale parameters of a probability distribution. The main advantage is that the linearity of the probability plot is a good measure of the adequacy of the distribution fit. The correlation coefficient between the points on the probability plot is a good measure of the linearity of the probability plot. It is an easy technique to implement for a wide variety of distributions with a single shape parameter. The basic requirement is to be able to compute the percent point function, which is needed in the computation. Probability plots provide insight of the sensitivity of shape parameter. If the plot is relatively flat in the neighborhood of the optimal value of the shape parameter, then it is a strong indication that the fitted model will not be sensitive to small deviations. The maximum correlation value provides a method for comparing the distributions as well as identifying the best value of the shape parameter for a given distribution. For example, one could use the probability plots for Weibull, log-normal, etc. Comparing the maximum correlation coefficient achieved for each distribution can help in selecting the best distribution to use. This method also has a few disadvantages. It is limited to distributions with a single shape parameter. Probability plots are widely available in common statistical software packages for a limited number of distributions. Significance levels for the correlation coefficient (i.e., if the maximum correlation value is above a given value, then the distribution provides an adequate fit for the data with a given confidence level) can be estimated only for a limited number of distributions, which is a serious limitation.

1.21 EXERCISES

1. If $A \Rightarrow B$ says that A is true, B is true; this is deductive logic, leading to the conclusion with some certainty. But in the policeman example, A represents events as {broken windows, etc.} and B represents events as {man is a thief}. This leads to the statement that if B is true, A becomes more plausible which is ok. But in the case of probability, there always exists a percent of doubt of any event happening. Then how does deductive logic become important in such cases (for self-reasoning)?
2. State an example, identifying the space of events and try to deduce the plausibility of occurrence of proposition A

Solution

On space, S, $m = 2n$ points, there are $2m$ functions. For example, for $n = 1$, we have four functions, namely:

A	T	F
$f_1(A)$	T	T
$f_2(A)$	T	F
$f_3(A)$	F	T
$f_4(A)$	F	F

The same shall also be written as

$$\begin{aligned} f_1(A) &= A \\ f_2(A) &= A + \bar{A} \\ f_3(A) &= \bar{A} \\ f_4(A) &= A\bar{A} \end{aligned}$$

3. If $P(A) = 3/5$ and $P(B) = 1/5$, find $P(A \cap B)$ if A and B are independent events

Solution

It is given that $P(A) = 3/5$ and $P(B) = 1/5$
A and B independent events. Therefore,

$$P(A \cap B) = P(A) * P(B) = \frac{3}{5} * \frac{1}{5} = \frac{3}{25}$$

4. Two cards are drawn at random and without replacement from a pack of 52 playing cards. Find the probability that both the cards are black.
There are 26 black cards in a deck of 52 cards
Let $P(A)$ be the probability of getting a black cards on the first draw

$$P(A) = \frac{26}{52} = \frac{1}{2}$$

Let $P(B)$ be the probability of getting a black card on the second draw. Since the card is not replaced

$$P(B) = \frac{25}{51}$$

Thus the probability of getting the cards black

$$\frac{1}{2} * \frac{25}{52} = \frac{25}{104}$$

5. A set of concrete cubes of an offshore deck need to be ascertained for quality assurance. Three randomly selected cubes are to be tested under universal testing machine (UTM), based on which quality certification (QC) will be issued. If all three cubes show the desired strength, then the construction of the deck slab of an offshore platform is approved for payment and considered to be passing the QC satisfactorily; otherwise it would be rejected. Find the probability that a set containing 15 cubes out of which only 12 are of good quality would yield a positive QC.

Solution

Let A, B, and C be the respective events that the first, second, and third tested cubes are good quality.

Therefore, the probability is that the first tested cube is good quality, $P(A) = 12/15$. As the cubes are not replaced (as the test destructive), the probability of getting the second cube of good quality is $P(B) = 11/14$. Similarly, the probability of getting the third cube of good quality, $P(C) = 10/13$. It is also noted that the quality check will be declared positive if all the three cubes tested would be assessed as good quality. Thus, the probability of getting all the cubes of good quality (12/15)*11/14)*10/13) = 44/91. Therefore, the probability that the quality check is positive is 44/91.

6. A coin and an unbiased die are tossed. Show that they are independent.

Solution

Sample space S is given by

$$S = \begin{Bmatrix} (H,1)*(H,2)*(H,3)*(H,4)*(H,5)*(H,6) \\ (T,1)*(T,2)*(T,3)*(T,4)*(T,5)*(T,6) \end{Bmatrix}$$

Let A: head appears on the coin

$$A = \{(H,1)*(H,2)*(H,3)*(H,4)*(H,5)*(H,6)\}$$

$$\Rightarrow P(A) = \frac{6}{12} = \frac{1}{2}$$

Let B: number 3 appearing on the die $\{(H,3) * (T,3)\}$

$$P(B) = \frac{2}{12} = \frac{1}{6}$$

$$A \cap B = \{(H,3)\}$$

$$P(A \cap B) = \frac{1}{12}$$

$$P(A) * P(B) = \frac{1}{2} * \frac{1}{6} = P(A \cap B)$$

Therefore, A and B are independent events.

7. In a lucky draw, numbers marked from 1 to 6 are rolled and placed in an urn. Let A be the event, when the lucky draw shows an even number and B shows an odd number. Check whether both these events are independent. When a luck draw is to be made, the sample space (S) governing the draw is

$$S = \{1, 2, 3, 4, 5, 6\}$$

Let A: the number is even $=\{2,4,6\}$

$$\Rightarrow P(A)=\frac{3}{6}=\frac{1}{2}$$

B: the number is odd $=\{1,2,3\}$

$$\Rightarrow P(B)=\frac{3}{6}=\frac{1}{3}$$

$$A\cap B=\{2\}$$

$$P(AB)=P(A\cap B)=\frac{1}{6}$$

$$P(A)*P(B)=\frac{1}{2}*\frac{1}{2}=\frac{1}{4}\neq\frac{1}{6}$$

$$\Rightarrow P(A)*P(B)\neq P(AB)$$

Therefore, A and B are not independent.

8. If $P(A) = 1/4$, $P(A \cup B) = 3/5$, and $P(B) = p$, find p if given that the events A and B are such that they are (i) mutually exclusive (ii) independent.

Solution

$$P(A)=\frac{1}{4},\ P(A\cap B)=\frac{3}{5},\ \text{and } P(B)=p$$

i. When A and B are mutually exclusive $A\cap B=0$

$$P(A\cap B)=0$$

$$P(A\cup B)=P(A)+P(B)-P(A\cap B)$$

$$\Rightarrow\frac{3}{5}=\frac{1}{4}+p-0$$

$$\Rightarrow p=\frac{3}{5}-\frac{1}{4}=\frac{7}{20}$$

ii. when A and B are independent

$$P(A\cap B)=P(A)*P(B)=\frac{1}{2}p$$

$$P(A \cup B) = P(A) + P(B) - P(A \cap B)$$

$$\Rightarrow \frac{3}{5} = \frac{1}{4} + p - \frac{1}{2}p$$

$$\Rightarrow \frac{3}{5} = \frac{1}{4} + \frac{p}{2}$$

$$\Rightarrow \frac{p}{2} = \frac{3}{5} - \frac{1}{4} = \frac{7}{20}$$

$$\Rightarrow p = \frac{7}{10}$$

9. Suppose an offshore platform is inspected for its service life estimate. A few damages are visually observed by the team of experts. The column members of the platform are damaged by impact of vessels/boats. The deck of the platform is also seen to be damaged by dropped objects from the crane in operation. Considering various combinations of these effects, one is interested to know their consequences in the reliability estimate of the platform. Over a period of time, a series of observations are recorded namely: column damage is 0.3 and deck damage is 0.4. Considering both the observations are independent of each other, it is necessary to estimate the following:
   a. Determine the probability of common factors that can influence both the observed damages
   b. Determine the probability of factors that can either cause damage to the column member or to the deck
   c. Determine the status of knowledge on the column member given that deck is damaged by dropped objects
   d. Determine the status of knowledge of deck given that column member is damaged by vessel impact

Solution

Let A and B be independent events with $P(A) = 0.3$ and $P(B) = 0.4$. We need to determine (i) $P(A \cap B)$; (ii) $P(A \cup B)$; (iii) $P(A \mid B)$; (iv) $P(B \mid A)$ as required.

It is given that $P(A) = 0.3$ and $P(B) = 0.4$

i. If A and B are independent events, then

$$P(A \cap B) = P(A) * P(B) = 0.3 * 0.4 = 0.12$$

ii. $P(A \cup B) = P(A) + P(B) - P(A \cap B)$

iii. $\Rightarrow P(A \cup B) = 0.3 + 0.4 - 0.12 = 0.58$ $P(A|B) = \dfrac{P(A \cap B)}{P(B)}$

$$\Rightarrow P(A|B) = \frac{0.12}{0.4} = 0.3$$

$$P(B|A) = \frac{P(A \cap B)}{P(A)} \qquad \Rightarrow P(B|A) = \frac{0.12}{0.3} = 0.4$$

10. In the probability analyses, if A and B are two events such that $P(A) = 1/4$, $P(B) = 1/2$, and $P(A \cap B) = 1/8$,
determine P (not A and not B)

Solution

$$P(b) = \frac{1}{2} \text{ and } P(A \cap B) = \frac{1}{8}$$

It is given that
P(not on A and not on B) $= P(A' \cap B')$
P(not on A and not on B) $= P(A \cup B)'[A' \cap B' = (A \cup B)']$

$$= 1 - P(A \cup B)$$

$$= 1 - [P(A) + P(B) - P(A \cap B)]$$

$$= 1 - \left[\frac{1}{4} + \frac{1}{2} - \frac{1}{8}\right]$$

$$= 1 - \frac{5}{8}$$

$$= \frac{3}{8}$$

11. Events A and B are such that $P(A) = 1/2$, $P(B) = 7/12$, and P(not A or not B) $=1/4$. State whether A and B are independent events?
It is given that $P(A) = 1/2$, $P(B) = 7/12$, and P(not A or not B) $= 1/4$

$$\Rightarrow P(A' \cup B') = \frac{1}{4}$$

$$\Rightarrow P((A \cap B)') = \frac{1}{4}$$

$$\left[A' \cup B' = (A \cap B)'\right]$$

$$\Rightarrow 1 - P(A \cap B) = \frac{1}{4}$$

$$\Rightarrow P(A \cap B) = \frac{3}{4}$$

However,

$$P(A) * P(B) = \frac{1}{2} * \frac{7}{12} = \frac{7}{24}$$

Here 3/4 ≠ 7/24

$$P(A \cap B) \neq P(A) * P(B)$$

Therefore, *A* and *B* are independent events.

12. Based on visual observation, it is reported that the column members of a jacket platform are found to be damaged by vessel impact (considered as event *A*). The deck of the platform is also found to be damaged by dropped objects from the crane during the operation (considered as event *B*). Considering various combinations of these effects, one is interested to know their consequences in the reliability estimate of the platform. Considering that these two events are independent of each other such that their probabilities are 0.3 and 0.6, respectively, determine the following
    i. *P*(*A* and *B*)
    ii. *P*(*A* and not *B*)
    iii. *P*(*A* or *B*)
    iv. *P*(neither *A* nor *B*)

    It is given that *P*(*A*)= 0.3 and *P*(*B*) = 0.6
    Also, *A* and *B* are independent events
    i. *P*(*A* and *B*) = *P*(*A*).*P*(*B*)
       $\Rightarrow P(A \cap B) = 0.3*0.6 = 0.18$
    ii. *P*(*A* and not *B*) = $P(A \cap B')$
       $= P(A) - P(A \cap B)$
       = 0.3–0.18
       = 0.12
    iii. *P*(*A* or *B*) = $P(A \cup B)$
       $= P(A) + P(B) - P(A \cap B)$
       = 0.3 + 0.6–0.18
       = 0.72
    iv. *P*(neither *A* nor *B*) = $P(A' \cap B')$
       $= P((A \cup B)')$
       $= 1 - P(A \cup B)$
       = 1–0.72
       = 0.28
13. The probability of solving specific problem independently by *A* and *B* are 1/2 and 1/3, respectively. If both try to solve the problem independently, find the probability that the problem is solved.

    Solution

    Probability of solving the problem by *A*, *P*(*A*) = 1/2
    Probability of solving the problem by *B*, *P*(*B*) = 1/3

Since the problem is solved independently by A and B,

$$P(AB) = P(A) * P(B) = \frac{1}{2} * \frac{1}{3} = \frac{1}{6}$$

Probability that A cannot solve the problem independently is

$$P(A') = 1 - P(A) = 1 - \frac{1}{2} = \frac{1}{2}$$

Similarly for B, it is given by

$$P(B') = 1 - P(B) = 1 - \frac{1}{3} = \frac{2}{3}$$

The probability that the problem is solved = $P(A \cup B)$
= $P(A) + P(B) - P(AB) = 0.66$.

14. In a problem related to the hike in a worker's salary at a construction site, the following data are observed. Demand in hike in the salary is due to (i) increase in cost of community living as the construction site is located downtown; and (ii) increase in municipal charges of water and electricity. On a statistical survey done by the management team, it was found that 60% of the employees state community living as the prime reason, 40% state the increase in municipal taxes and the remaining state both the reasons. One of the persons from the group is chosen to assess the correct factor at random as this would decide the % hike in the salary as per the local government norms.
    a. Find the probability that the chosen person supports either of the reasons.
    b. If the chosen person supports the first reason, find the probability that he also supports the second one.
    c. If he supports the second reason, find the probability that he also supports the first reason.

Solution

Let A denote the population supporting increase in the cost of community living and B denote the population supporting the increasing the municipal charges. It is given that

$$P(A) = 60\% = \frac{60}{100} = \frac{3}{5}$$

$$P(B) = 40\% = \frac{40}{100} = \frac{2}{5}$$

$$P(A \cap B) = 20\% = \frac{20}{100} = \frac{1}{5}$$

i. Probability that the chosen person supports either of the reason is given by

$$(A \cup B)' = 1 - P(A \cup B)$$

$$= 1 - \{P(A) + P(B) - P(A \cap B)\}$$

$$= 1 - \left(\frac{3}{5} + \frac{2}{5} - \frac{1}{5}\right) = \frac{1}{5}$$

ii. Probability that a randomly chosen person supports first reason, if he supports the second reason is given by $P(B \mid A)$

$$P(B|A) = \frac{P(B \cap A)}{P(A)}$$

$$= \frac{1/5}{3/5}$$

$$= \frac{1}{3}$$

iii. Probability that a randomly chosen person supports second reason, if he supports the first one is given by $P(A \mid B)$

$$P(A|B) = \frac{P(A \cap B)}{P(B)}$$

$$= \frac{1/5}{2/5}$$

$$= \frac{1}{2}$$

15. An urn contains 5 red and 5 black balls. A ball is drawn at random, its color is noted and is returned to the urn. Moreover, two additional balls of the color drawn are put in the urn and then a ball is drawn at random. What is the probability that the second ball is red?

Solution

The urn contains 5 red and 5 black balls
Let a red ball be drawn in the first attempt
P(drawing a red ball) = 5/10 = 1/2
If two red balls are added to the urn, then the urn contains 7 red and 5 black balls
P(drawing a red ball) = 7/12
Let a black ball be drawn in the first attempt

P(drawing a black ball in the first attempt) = 5/10 = 1/2
If two black balls are added to the urn, then the urn contains 5 red and 7 black balls
P (drawing a red ball) = 5/12
Therefore, the probability of drawing a second ball as red is

$$= \frac{1}{2} * \frac{7}{12} + \frac{1}{2} * \frac{5}{12} = \frac{1}{2}\left(\frac{7}{12} + \frac{5}{12}\right) = \frac{1}{2} * 1 = \frac{1}{2}$$

16. A bag contains 4 red and 4 black balls, another bag contains 2 red and 6 black balls. One of the two bags is selected at random and a ball is drawn from the bag which is found to be red. Find the probability that the ball is drawn from first bag.

Solution

Let E_1 and E_2 be the events of selecting bag and second bag, respectively

$$P(E_1) = P(E_2) = \frac{1}{2}$$

Let A be the event of getting a red ball

$$\Rightarrow P\ (A|E_1) = P(\text{drawing a red ball from first bag}) = \frac{4}{8} = \frac{1}{2}$$

$$\Rightarrow P\left(\frac{A}{E_2}\right) = P(\text{drawing a red ball from second bag}) = \frac{2}{8} = \frac{1}{4}$$

The probability of drawing a ball from the first bag, given that it is red, is given by $P(E_2/A)$.
By using Bayes' theorem, we obtain

$$P\left(\frac{E_1}{A}\right) = \frac{P(E_1) * P(A/E_1)}{P(E_1) * P(A/E_1) + P(E_2) * P(A/E_2)}$$

$$= \frac{(1/2) * (1/2)}{(1/2) * (1/2) + (1/2) * (1/4)}$$

$$= \frac{1/4}{(1/4) + (1/8)}$$

$$= \frac{1/4}{3/8}$$

$$= \frac{2}{3}$$

17. Of the students in a college, it is known that 60% reside in the hostel and 40% are day scholars (not residing in the hostel). The previous year's results report that 30% of all students who reside in the hostel attain an A grade and 20% of day scholars attain an A grade in their annual examination. At the end of the year, one student who secured an A grade is chosen at random from the college. What is the probability that the student resides in the hostel?

Solution

Let E_1 and E_2 be the events that the student resides in the hostel and is a day scholar, respectively, and A be the event that the chosen student gets grade A

$$P(E_1) = 60\% = \frac{60}{100} = 0.6$$

$$P(E_2) = 40\% = \frac{40}{100} = 0.4$$

$$P\left(\frac{A}{E_1}\right) = P(\text{student getting an A grade resides in the hostel}) = 30\% = 0.3$$

$$P\left(\frac{A}{E_2}\right) = P(\text{student getting an A grade resides in the hostel}) = 20\% = 0.2$$

The probability that a randomly chosen student resides in the hostel, given that he has an A grade, is given by $P(E_1/A)$.
By using Bayes' theorem, we obtain

$$P\left(\frac{E_1}{A}\right) = \frac{P(E_1) * P(A/E_1)}{P(E_1) * P(A/E_1) + P(E_2) * P(A/E_2)}$$

$$= \frac{0.6 * 0.3}{0.6 * 0.3 + 0.4 * 0.2}$$

$$= \frac{0.18}{0.26}$$

$$= \frac{18}{26}$$

$$= \frac{9}{13}$$

18. In answering a question on a multiple choice test, a student either knows the answer or guesses. Let 3/4 be the probability that he knows the answer and 1/4 be the probability that he guesses. Assume that a student who guesses the answer gets the correct answer with a probability 1/4. What is the probability that the student knows the answer given that he answered correctly?

Solution

Let E_1 and E_2 be the respective events that the student knows the answer and he guesses the answer.
Let A be the event that the answer is correct

$$P(E_1) = \frac{3}{4}$$

$$P(E_2) = \frac{1}{4}$$

The probability that the student answered correctly, given that he knows the answer, is 1.

The probability that the student answered correctly, given that he knows the answer, is 1/4 $P(A/E_2) = 1/4$.

The probability that the student knows the answer, given that he answered correctly, is given by $P(E_1/A)$.
By using Bayes' theorem we obtain

$$P\left(\frac{E_1}{A}\right) = \frac{P(E_1) * P(A/E_1)}{P(E_1) * P(A/E_1) + P(E_2) * P(A/E_2)}$$

$$= \frac{(3/4) * 1}{(3/4) * 1 + (1/4) + (1/4)}$$

$$= \frac{3/4}{(3/4) + (1/16)}$$

$$= \frac{3/4}{13/16}$$

$$= \frac{12}{13}$$

19. A nondestructive test method is 99% effective in detecting the crack initiation when it is present. However, the test also yields false-positive results for 0.5% of the structures tested earlier. This implies that if a good structure is tested then with probability the test will imply that the structure has a defect is 0.005. If 0.1% of the structures are actually deficient, what is the probability that the structure is truly deficient given that the test result from the non destructive testing (NDT) is positive?

Solution

Let E_1 and E_2 be the respective events that the structure is truly deficient or not.
Since E_1 and E_2 are events complimentary to each other, the following holds good:

$$P(E_1) + P(E_2) = 1$$

$$\Rightarrow P(E_2) = 1 - P(E_1) = 1 - 0.001 = 0.999$$

Let A be the event that the NDT result is positive

$$P(E_1) = 0.1\% = \frac{0.1}{100} = 0.001$$

$$P\left(\frac{A}{E_1}\right) = P(\text{NDT result is positive and the structure is truly deficient}) = 99\% = 0.99$$

$$P\left(\frac{A}{E_2}\right) = P(\text{NDT result is positive but the structure is not deficient}) = 0.5\% = 0.005$$

Probability that the structure is truly deficient and NDT result is also positive is given using Bayes' theorem:

$$P\left(\frac{E_1}{A}\right) = \frac{P(E_1) * P(A/E_1)}{P(E_1) * P(A/E_1) + P(E_2) * P(A/E_2)}$$

$$= \frac{0.001 * 0.99}{0.001 * 0.99 + 0.99 * 0.005}$$

$$= \frac{22}{133}$$

20. There are three coins. One is a two-headed coin (having head on both faces), another is a biased coin that comes up heads 75% of the time and third is an unbiased coin. One of the three coins is chosen at random and tossed and it shows head. What is the probability that it was the two-headed coin?
Let E_1 and E_2 be the respective events of choosing a two-headed coin, a biased coin, and an unbiased coin.

$$P(E_1) = P(E_2) = P(E_3) = \frac{1}{3}$$

Let A be the event that the coin shows heads
A two-headed coin will always show heads

$$P\left(\frac{A}{E_1}\right) = P(\text{coins showing heads, given that it is a two-headed coin}) = 1$$

Probability of heads coming up given that it is a biased coin = 75%

$$P\left(\frac{A}{E_2}\right) = P(\text{coins showing heads, given that it is a two-headed coin}) = \frac{75}{100} = \frac{3}{4}$$

Since the third coin is unbiased, the probability that it shows heads is always = 1/2

$$P\left(\frac{A}{E_3}\right) = P(\text{coins showing heads, given that it is an unbiased coin}) = \frac{1}{2}$$

The probability that the coin is two headed, given that it shows heads, is given by

$P(E_1/A)$ by using Bayes' theorem. We obtain

$$P\left(\frac{E_1}{A}\right) = \frac{P(E_1) * P(A/E_1)}{P(E_1) * P(A/E_1) + P(E_2) * P(A/E_2) + P(E_3) * P(A/E_3)}$$

$$= \frac{(1/3) * 1}{(1/3) * 1 + (1/3) * (3/4) + (1/3) * (1/2)}$$

$$= \frac{1/3}{1/3(1 + (3/4) + (1/2))}$$

$$= \frac{1}{9/4}$$

$$= \frac{4}{9}$$

21. An insurance company insured 2000 scooter drivers, 4000 car drivers, and 6000 truck drivers. The probability of accidents are 0.01, 0.03, and 0.15, respectively. One of the insured persons has with an accident. What is the probability that he is a scooter driver?

Let E_1, E_2, and E_3 be the respective events that the driver is a scooter driver, a car driver, and a truck driver
Let A be the event that the person meets with an accident

There are 2000 scooter drivers, 4000 car drivers, and 6000 truck drivers
Total number of drivers = 2000 + 4000 + 6000 = 12,000

$$P(E_1) = P(\text{driver is a scooter driver}) = \frac{2000}{12,000} = \frac{1}{6}$$

$$P(E_2) = P(\text{driver is a car driver}) = \frac{4000}{12,000} = \frac{1}{3}$$

$$P(E_3) = P(\text{driver is a truck driver}) = \frac{6000}{12,000} = \frac{1}{2}$$

$$P\left(\frac{A}{E_1}\right) = P(\text{scooter driver met with an accident}) = 0.01$$

$$P\left(\frac{A}{E_2}\right) = P(\text{car driver met with an accident}) = 0.03$$

$$P\left(\frac{A}{E_3}\right) = P(\text{truck driver met with an accident}) = 0.15$$

The probability that the driver is a scooter driver given that he met with an accident is given by $P(E_1/A)$. By using Bayes' theorem. We obtain

$$P\left(\frac{E_1}{A}\right) = \frac{P(E_1) * P(A/E_1)}{P(E_1) * P(A/E_1) + P(E_2) * P(A/E_2) + P(E_3) * P(A/E_3)}$$

$$= \frac{(1/6) * (1/100)}{(1/6) * (1/100) + (1/3) * (3/100) + (1/2) * (15/100)}$$

$$= \frac{(1/5) * (1/100)}{(1/100)\left(1/6 + 1 + 15/2\right)}$$

$$= \frac{1/6}{104/2}$$

$$= \frac{1}{6} * \frac{12}{104}$$

$$= \frac{1}{52}$$

22. A factory has two machines A and B. Past records show that machine A produced 60% of the items of output and machine B produced 40% of the items. Further, 2% of the items produced by machine A and 1% production

by machine B were defective. All the items are put into one stockpile and then one item is chosen at random from this and is found to be defective. What is the probability that it was produced by machine B?

Let E_1 and E_2 be the respective events of items produced by machines A and B. Let X be the event that the produced item was found to be defective. Probability of items produced by machine A, $P(E_1) = 60\% = 3/4$
Probability of items produced by machine B, $P(E_1) = 40\% = 2/5$

Probability that machine A produced defective items, $P(X/E_1) = 2\% = 2/100$. Probability that machine B produced defective items, $P(X/E_2) = 1\% = 1/100$. The probability that the randomly selected item was from machine B, given that it is defective is given by $P(E_2/X)$. By using Bayes' theorem, we obtain

$$P\left(\frac{E_2}{A}\right) = \frac{P(E_2) * P(A/E_2)}{P(E_1) * P(A/E_1) + P(E_2) * P(A/E_2)}$$

$$= \frac{(2/5) * (1/100)}{(3/5) * (2/100) + (2/5) * (1/100)}$$

$$= \frac{2/500}{(6/500) + (2/500)}$$

$$= \frac{2}{8}$$

$$= \frac{1}{4}$$

23. Two groups are competing for the position on the board of directors of a corporation. The probability that the first and the second groups will win are 0.6 and 0.4, respectively. Further, if the first group wins, the probability of introducing a new product is 0.7 and the corresponding probability is 0.3 if the second group wins. Find the probability that the new product was introduced by the second group.
Let E_1 and E_2 be the respective events that the first and second group win the competition. Let A be the event of introducing a new product.

$$P(E_1) = \text{probability that the first group wins the competition} = 0.6$$

$$P(E_2) = \text{probability that the second group wins the competition} = 0.4$$

$$P\left(\frac{A}{E_1}\right) = \text{probability of introducing a new product if the first group wins} = 0.7$$

$$P\left(\frac{A}{E_2}\right) = \text{probability of introducing a new product if the second group wins} = 0.3$$

The probability that the new product is introduced by the second group is given by $P(A/E_2)$
By using Bayes' theorem, we obtain

$$P\left(\frac{E_2}{A}\right) = \frac{P(E_2) * P(A/E_2)}{P(E_1) * P(A/E_1) + P(E_2) * P(A/E_2)}$$

$$= \frac{0.4 * 0.3}{0.6 * 0.7 + 0.4 * 0.3}$$

$$= \frac{0.12}{0.54}$$

$$= \frac{2}{9}$$

24. A manufacturer has three machine operators A, B, and C. The first operator A produces 1% defective items, whereas the other two operators B and C produce 5% and 7% defective items, respectively. A is on the job for 50% of the time, B is on the job for 30% of the time, and C is on the job for 20% of the time. A defective item is produced, what is the probability that was produced by A?

Let E_1, E_2, and E_3 be the respective events of the time consumed by machines A,B, and C for the job.

$$P(E_1) = 50\% = \frac{50}{100} = \frac{1}{2}$$

$$P(E_2) = 30\% = \frac{30}{100} = \frac{3}{10}$$

$$P(E_3) = 20\% = \frac{20}{100} = \frac{1}{5}$$

Let X be the event of producing defective items

$$P\left(\frac{X}{E_1}\right) = 1\% = \frac{1}{100}$$

$$P\left(\frac{X}{E_2}\right) = 5\% = \frac{5}{100}$$

$$P\left(\frac{X}{E_3}\right) = 7\% = \frac{7}{100}$$

The probability that the defective item was produced by A is given by $P(E_1/X)$. By using Bayes' theorem, we obtain

$$P\left(\frac{E_1}{X}\right) = \frac{P(E_1) * P(X/E_1)}{P(E_1) * P(X/E_1) + P(E_2) * P((XE_2)) + P(E_3) * P(X/E_3)}$$

$$= \frac{(1/2) * (1/100)}{(1/2) * (1/100) + (3/10) * (5/100) + (1/5) * (7/100)}$$

$$= \frac{1/2}{17/5}$$

$$= \frac{(1/100) * (1/2)}{(1/100)\left((1/2) + (3/2) + (7/5)\right)}$$

$$= \frac{5}{34}$$

25. If A and B are two events such that $A \subset B$ and $P(B) \neq 0$, then which of the following is correct?
    A. $P(AB) = P(B)/P(A)$
    B. $P(AB) < P(A)$
    C. $P(AB) \geq P(A)$
    D. None of this

    If $A \subset B$, then $A \cap B = A$
    $\Rightarrow P(A \cap B) = P(A)$
    Also , $P(A) < P(B)$
    Considering the following

$$P(A \mid B) = \frac{P(A \cap B)}{P(B)} = \frac{P(A)}{P(B)} \neq \frac{P(B)}{P(A)}$$

Considering the following

$$P(A \mid B) = \frac{P(A \cap B)}{P(B)} = \frac{P(A)}{P(B)}$$

It is known that $P(B) \leq 1$

$$\Rightarrow \frac{1}{P(B)} \geq 1$$

$$\Rightarrow \frac{P(A)}{P(B)} \geq P(A)$$

We also know that

$$\Rightarrow P(A \mid B) \geq P(A)$$

26. $P(A \mid B)$ is not less than $P(A)$
27. Thus, statement C is correct.

2 Structural Reliability Theory

2.1 RELIABILITY

The failure of offshore structures is a consequence of decisions made under uncertain conditions. This includes different type of failures such as temporary failures causing shutdown of the platform, maintenance failures resulting in temporary uphold of oil and gas production, failures in design causing decrease in payload or insufficient operational conditions, etc., which need to be addressed. In all such cases, the definition of failure becomes important, which is generally expressed in terms of the probability of failure. This is assessed as *inability to perform its intended function adequately on demand for a period of time under specific conditions*. The converse of probability of failure is reliability and is defined in terms of the success of a system. Reliability of a system is the probability of a system performing its required function adequately for specified period of time under stated conditions. In dealing with design, uncertainties are unavoidable as seen in Chapter 1. Uncertainties, as applicable to offshore structure design are classified into two broad types, namely: (a) those associated with natural randomness; and (b) those associated with inaccuracies in the prediction and estimation of reality. The former type is known as *aleatory type*, which arises out of capacity estimates (uncertainties in loads) whereas the latter is known as *epistemic type*.

As the question of uncertainty and randomness of data is central to design and analysis, decisions are made on the basis of information, which is either limited or incomplete. It becomes therefore necessary to supplement the traditional deterministic methods of analysis using statistics and probability. Practical experiences, applied with engineering judgment are vital in the design of offshore structures. Let us recall that uncertainties can be classified as: (i) those arising from materials (e.g., modulus of concrete, steel, etc.); (ii) those arising from loads (e.g., random loads); and (iii) those arising from mathematical modeling and analysis methods (Borgman et al., 1982). A reliability approach is often used in the framework of quality control. A reliability-based approach to design, which involves the occurrence of rare events, does not allow feedback on the part or product in question. It is therefore not a prediction, which is validated by statistical control. Risk in the design procedure is based primarily on the statistical knowledge of the elementary variables used in the modeling. Knowledge of material properties, geometry, boundary conditions, and actions are also other variables, which are equally important. Combination of the elementary variables composes a deterministic but complex model, though variables are probabilistic in nature. Mechanical behavior of structures cannot, in general, be studied using the statistics of failure. Failure is therefore a simple combination of complex components rather than a complex combination of simple components. The latter case is more frequently encountered when dealing with system reliability. Should all the failure modes be possible in the design, even then relations between the elementary variables, components, and system are required to be examined. Such preliminary analysis becomes a vital step as the identification of potential failure implies the implementations of measures for minimizing risks, even without the reliability calculations. There is no use in devoting considerable effort to a well-identified situation if a critical situation has been overlooked. Theoretical reliability is therefore calculated based on the ideal representation of data and models, but practical reliability is used to address quality control. Table 2.1 briefly presents their features.

Reliability methods separate the situations that the designer decides to consider acceptable or otherwise; decision are made in a binary manner. Alternatively, it can be associated by increasing the degree-of-satisfaction, progressively from a failure situation to a safe operating condition (CEB Bulletin, 1976). The failure scenario is represented by one (or more) performance function(s) that delimit both the domains, namely: (i) safety domain, in which the performance function takes positive values; and (ii) failure domain, in which it takes negative or zero values (Chen and Lind, 1983; Coppola, 1984). Reliability analyses require, in addition, a failure scenario, as shown in Figure 2.1.

TABLE 2.1
Practical and Theoretical Reliability

In design	Analyzing all the possible failure modes that are elementary variables, components, and systems; taking steps to minimize the risks
In construction and maintenance	Quality assurance verifies the assumptions underlying the design
Practical reliability = Quality assurance and quality control	
Theoretical reliability is conditioned by practical reliability	
Theoretical reliability = Index or a measure	

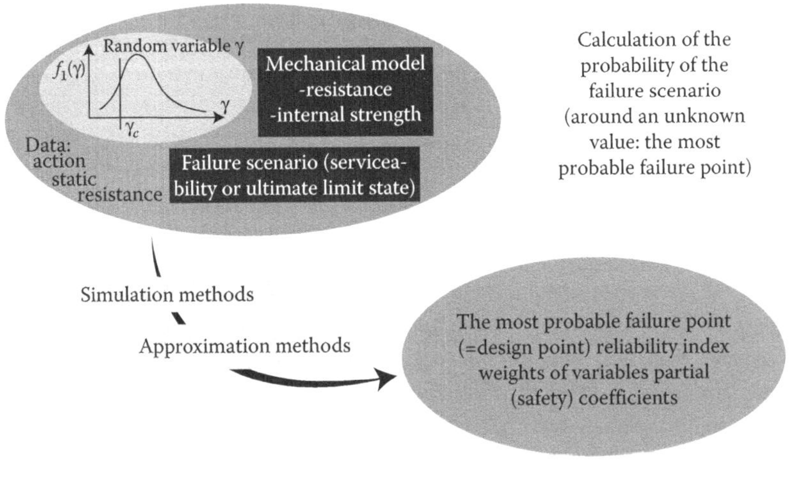

FIGURE 2.1 Components of reliability analysis.

Computing the probability of failure scenario around an unknown value is the most vital part of reliability analyses. Mathematically, estimating the reliability index (β) on the basis of fixing the design point in the failure domain is very crucial. This will lead to estimating a limit-state function, which is a zero-performance function. This binds two independent functions namely: (i) one for internal strength offered by the material; and (ii) other for evaluation of resistances offered by the structural form. For example, in offshore compliant structures like tension leg platforms (TLPs), where structural form controls the design, it is important to know the resistance offered by the structural form (or) weakness offered by the external loads. The random variable, which is used to model the uncertainty, is the marginal difference between the internal strength and resistance. In statistical sense, the main objective of the reliability analysis is to evaluate the probability that this margin has a positive value. Based on the data pertaining to the external forces, structural form and its resistance are expressed in an appropriate reliability model. The main objective is therefore to calculate a reliability index. In addition to this, the most probable failure point (design point), the sensitivity factors responsible for the anticipated failure and evaluation of partial (safety) coefficients are also important. Thus, the analytical procedure of reliability modeling excites the internal strength and resistance models by choice of the data to produce the required results. In the case of offshore structures, there are two outcomes possible; either success or failure. Both are mutually exclusive and exhaustive. This is due to the fact that both them are dependent on a variety of factors. As no alternate outcome (other than success or failure) is feasible, the following statement is valid on the basis of probability theory as seen in Chapter 1:

$$P[\text{Success}] + P[\text{Failure}] = 1 \tag{2.1}$$

The probability of success of the structure is reliability is therefore given by

$$R = 1 - P[\text{Failure}] \tag{2.2}$$

where R is reliability. It is important to note that reliability is expressed in terms of confidence built-up of the system in a positive sense and not as an index of failure; this is one of the better ways of looking at a failure theory (Cornel, 1969; Denson, 1998).

2.2 VARIABLES IN RELIABILITY STUDY

The variables (choice) in a reliability study depend on the nature of the problem posed, but have high interaction between the various specialties that collectively provide additional information to the decision maker. There can be many examples quoted to support this statement. For example, the statistician analyzes the data, conditions them, and models them as a function of a predictable use. The probability engineer proposes the methods and tools for the calculation. The reliability engineer analyzes the failure modes, guarantees quality assurance, and defines the failure scenarios and their combinations. The structural engineer analyzes the response behavior of the materials and structures, and guarantees the proper use of the mechanical

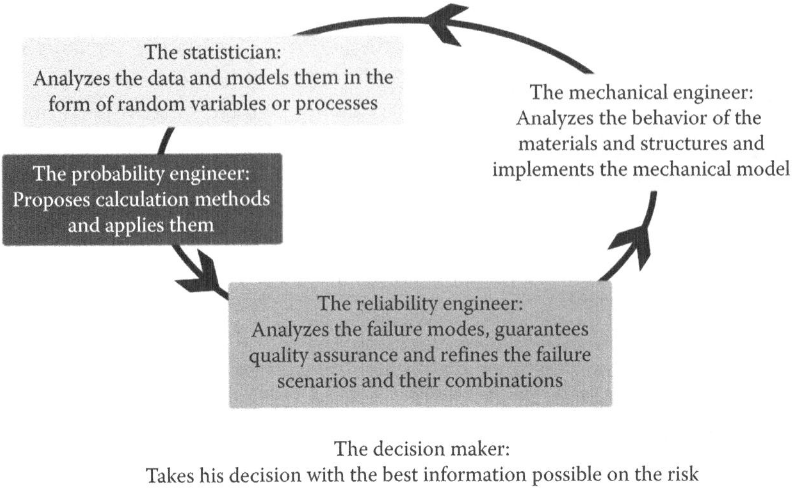

FIGURE 2.2 Interrelationship of variables in reliability study.

models used. Each one of them uses a variety of variables to examine the assigned task and gives a useful output. Figure 2.2 shows their interrelationship.

2.3 PROBABILISTIC APPROACH

Probabilistic modeling aims at a range of outcomes for the given input data. For this to be true, the system includes randomness of data and other uncertainties. A probabilistic approach aims at determining the probability (p) of an outcome amongst the many that may occur. Probability can be expressed either in a percentage varying from 0 to 100 or as a fraction varying between 0 and 1. In engineering practice, lower risk usually means higher cost. Hence, a common question asked is how safe is safe enough or what risk is acceptable? This is commonly referred as ALARP, which means that risk is kept as low as reasonably practical. The level of acceptable risk is determined by the regulatory agencies, which is based on the guidelines formulated by a number of researchers and engineers involved in risk assessment. They indicate that the incremental risk should not be significant compared to other risks and that the risks should be reduced to ALARP level. Annual probability of failure corresponds to an expected factor of safety $E(F)$, which is a variable expressed in terms of standard deviation of the factor of safety. If the factor of safety is assumed to be normally distributed, the reliability index (β) is expressed by

$$\beta = \frac{E(F) - 1}{\sigma_F} \tag{2.3}$$

Figure 2.3 shows the variations of the reliability index, as a measure failure probability, varying from high to hazardous.

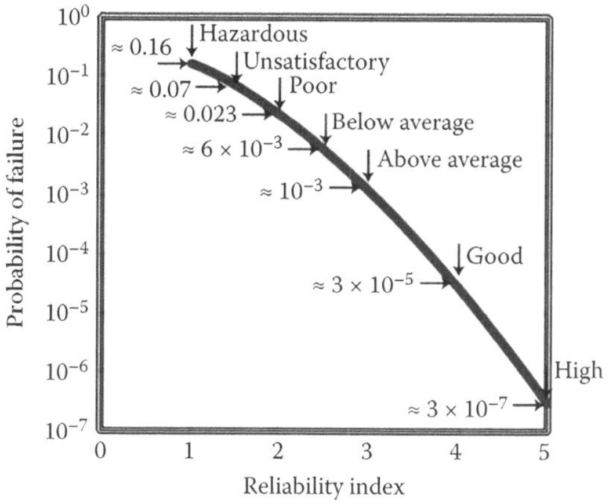

FIGURE 2.3 Relation between reliability index and probability of failure.

In the structural design of offshore structures, the design relies on deterministic analysis to arrive at appropriate cross-section dimensions and material to withstand the encountered loads. It is well known that material properties and the encountered loads in offshore structures are clear victims of uncertainties (Srinivasan Chandrasekaran, 2015a,b). Loads are highly uncertain due to their mean of occurrence and material and due to their significant degradation in a marine environment. Fortunately, variations that arise from the new geometric (structural) form of the offshore platforms are very marginal as this is usually based on the sound experimental investigations on the scaled models. An account of safety index is introduced in the design through partial safety factors for material properties and loads, but does not guarantee an absolute measure of failure probability (Ditlevsen, 1981; Freudenthal, 1947; Madsen, 1988; Madsen et al., 1986).

2.4 RELIABILITY LEVELS

There are different levels of reliability analysis, which can be used in any design methodology depending on the importance of the structure (Der Kiureghian et al., 1987; Ditlvesen, 1979). The term "level" is characterized by the extent of information about the problem that is used and provided. They can be grouped into four levels namely: level I, II, III, and IV, respectively. Grouping depends on the degree of sophistication applied to the treatment of the problem. In level I, appropriate characteristic values of random variables are chosen. They are used as partial safety factors in the design, which are estimated based on probabilistic considerations. Load and resistance factor design (LRFD) method falls under this category. In this level, the main objective is to minimize the error between the target value and the design deviate. Level II employs two values of each uncertain parameter namely mean and variance. They are also supplemented by the correlation between the parameters to improve the accuracy of the estimate. Level III methods encompass complete analysis of the problem. It involves integration of the multidimensional joint probability density function of the random variables, extended over the safety domain. Reliability is expressed in terms of the reliability index and failure probabilities. Level IV methods are appropriate for structures that are of major economic importance. They involve principles of engineering and economic analysis under uncertainty, and consider cost and benefit of construction, maintenance, repair, consequences of failure, interest on capital, etc. (Esary et al., 1967). Sensitive projects like offshore structures, nuclear power projects, transmission towers, highway bridges are suitable objects of level IV reliability studies. Although, level I methods are still most often used in structural design, level II methods are useful in evaluating the safety of a structure. This can also be used as a tool to determine the rational set of safety factors (Srinivasan Chandrasekaran, 2015b).

2.5 SPACE OF VARIABLES

The state of variables refer to characterization of the basic load and resistance parameters that are used to formulate the performance function. For "n" state variables, limit state function is represented by "n" parameters (Fiessler et al., 1979). If all loads (or load effects) are represented by the variable Q and total resistance (or capacity) by R, then the space of state variables is a two-dimensional space as shown in Figure 2.4. The boundary between both the domains is described by the limit state function $g(R,Q) = 0$. Since both R and Q are random variables, one can define a joint density function $f_{RQ}(r, q)$ and the probability of failure is calculated by integrating the joint density function over the failure domain (i.e., the region in which $g(R, Q) < 0$). Due to the complex shape of the failure domain that arises from the uncertainties of variables, this integration is difficult to evaluate; hence, the reliability index is used to quantify structural reliability.

2.6 ERROR ESTIMATION

Error departures from acceptable practice are an inevitable part of all analytical studies. They add a considerable degree-of-uncertainty to design and construction

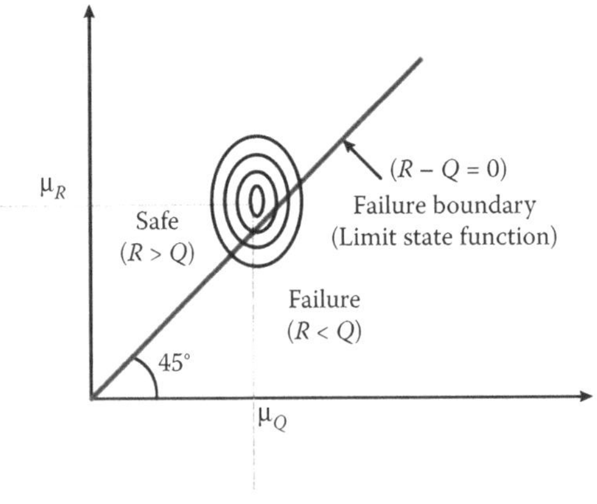

FIGURE 2.4 Space of variables marking safe and failure domains.

activities. In fact, surveys indicate that human errors are one of the main reasons of failure of major structures like buildings and bridges. As structural reliability is determined by error control, it is inevitable to minimize such errors. Errors are categorized according to their causes and consequences. There are two approaches to control errors, namely: (i) to reduce error frequency; and (ii) to minimize consequences.

2.6.1 Classification of Errors

The classification of errors on the basis of causes and consequences may be useful in the selection of efficient control measures. Analyses of causes are useful in identifying the occurrence mechanisms and can be planned to reduce the frequency of occurrence. Consequential errors are prevented by adopting special design procedures. Errors can be considered with regard to the person involved, phase of the building process, place, reason, and mechanism of occurrence. There are three fundamental types of errors, as discussed below:

1. *Errors of concept* are unintentional departures from the accepted practice due to insufficient knowledge.
2. *Errors of execution* are unintentional departures from what one believed to be an acceptable practice.
3. *Errors of intention* are intentional departure from the acceptable practice.

2.7 RELIABILITY AND QUALITY ASSURANCE

The reliability and quality assurance shall emphasize on the satisfactory performance of the intended function of the structure within the definite period. It involves responsibilities like setting targets for reliability and quality assurance, coordinating the related activities, preparing guidelines and documents, development and operation of the quality data reporting system, solution of quality and reliability problems at the structure and operability levels (Dowling, 1972; Freudenthal, 1947). Usually, a major part of the performance assurance is done at the design stage itself (level I reliability).

2.8 UNCERTAINTIES INHERENT IN DESIGN

Reliability analyses of offshore structures imply estimation of the limit state probabilities of a structure under adverse/environmental loading or its combination for its intended period of use (Grandt, 2004; Haldar and Mahadevan, 1995, 2000). As marine engineers understand the complexities that are present in loads and material characteristics, a proposed structural design is assessed for its safe and successful performance of the structure within its service life. Hence, safety is an important term used to quantify the satisfactory coverage of uncertainties in the design stage. *Safety* is related to an existing process, which has direct consequences to failure (IS 15656, 2006; Julian, 1957). While it is a design method but used only to assess the performance of a structure, the major advantage is that it is deterministic. Safety assessment methods give a close-form solution and therefore shall be applied to an existing structural system to assess its performance. On the other hand, *reliability* is a probability of realization of unsafeness (Benjamin and Cornel, 1970). It has a converse consequence of failure. It is probabilistic in approach and therefore used as one of the design methods (Ang and Tang, 1975; Lancaster, 2000). Accuracy of the results of the reliability approach essentially depends on the data based on which the results are arrived. Therefore, this has a probability of giving erroneous results if the data used is not reliable. Most importantly, reliability is assessed even before failure is foreseen and therefore reliability methods are accorded as a preventive forecast of failure (Chandrasekaran and Saha, 2011). Risk analysis of structures is an extension of reliability analysis to include the consequences of failure. The most important aspect of the reliability analysis is the consideration of uncertainties, which make structures vulnerable to failure for a predefined limit state. Accuracy of the reliability analysis depends upon how accurately all the uncertainties are accounted for in the analysis. First, it is practically impossible to identify all uncertainties; however, important ones can be identified. Second and most important, methods for modeling and analyzing them are not easy as some amount of uncertainty always remains associated with their modeling. Furthermore, analytical formulation of the limit state surface and integration of the probability density function within the domain of interest is complex; this may result in various approximations. As a result, different degrees of simplifications are made in the reliability analysis leading to the development of different reliability methods. Therefore, it is not possible to obtain the exact probability of failure of a structure for any event except for very simple ones.

Generally there are three types of uncertainties, which are dominant in reliability analysis namely: (a) randomness and variability of environmental loads; (b) statistical uncertainty, which arises due to estimation of parameters describing statistical models; and (c) model uncertainty, which arises due to imperfection of mathematical modeling of complex physical phenomena. Apart from these uncertainties, there are some others, which result from simplification of the problem at hand. For example, nonlinear analysis may be replaced by equivalent linear analysis; continuums may be represented by a discrete model with limited degrees of freedom, etc. Mostly, the uncertainty arising due to (a) is irreducible but those arising due to (b) and (c) can be reduced (Miller, 1981). For example, collection of more data or samples helps in providing better statistical parameters. Likewise, use of a more refined model may reduce the uncertainty due to (c). Other uncertainties, as mentioned above, may be reduced by performing more rigorous analysis with more sophisticated models of structures.

A unified approach for treating statistical and model uncertainties in reliability analysis is to use the Bayesian rule. This is used to update the model parameters and develop likelihood functions. With the help of these functions, posterior parameters or models are obtained from the prior ones, which are supposed to have less uncertainty as they are developed using more data and observations. Probabilistic models that are widely used to describe the distributions of different uncertain parameters are uniform distribution, extreme value distribution, lognormal distribution, and

Poisson distribution. The procedures for performing reliability analysis vary with the selection of the above models. When material and other uncertainties are introduced, the procedure for analysis may significantly differ (Rackwitz and Fiessler, 1976). For example, stochastic finite element model (FEM) analysis is used for random loading while, material and other uncertainties may be included by simple procedures in an approximate manner. In fact, various levels of approximations are often used to simplify the reliability analysis procedure consistent with the desired accuracy.

2.9 UNCERTAINTIES IN SYSTEM DESIGN OF OFFSHORE STRUCTURES

Uncertainties arise from various sources in offshore structures. Considering the fabrication process of the platform, uncertainties are bound to occur from nonavailability of the requisite material at the required time in the sequence of fabrication. An equivalent substitution may cause a minor uncertainty, which can be handled in the design stage itself (Shah et al., 2005; Srinivasan Chandrasekaran and Kiran, 2014a, b). In case of detailing in design, errors in detailing are generally noticed during reviews. Fabrication errors are captured during inspection and subsequently rectified. Uncertainties are quite complex in estimating stress concentration factors for fatigue analysis. Empirical rules are used for multi-planar joints, which do not represent true behavior. In the topside installation, major uncertainty is in lifting if the design does not match the lifting arrangement or if the detailing is inadequate to properly transfer lifting loads to balance the structure. These aspects are rigorously checked during the design review process. Uncertainties also arise during installation. They may be due to rough weather, which increases the loads during installation. In particular, uncertainties are more serious in case of the commissioning of large compliant platforms. It is interesting to note that the damaged compartment scenario is studied in the design stage to account for uncertainty that arises during sinking of the jacket after launching. Uncertainties also arise during grouting of shear keys, in particular. Alternatively, such issues are taken care of in the design by admitting a reduced factor of safety because remedial action is not possible for some reason (Srinivasan Chandrasekaran and Subarata, 2012). As a part of the design process, the method of safety checking also introduces uncertainties. Human error is the cause of most accidents and mishaps. This is due to inadequate knowledge, inadequate training and experience, lack of application and nonconformance to safety practices. While the main design work is actually checking the assumed member sizes for adequacy to carry the forces, which are well defined by standards codes, member adequacy checks are reasonably accurate with little uncertainty. However, joint adequacy checks are based on empirical formulae derived from model tests, as determined by specifications; these have a lot of uncertainties. These have been updated several times based on research and analysis. Even with the above list of uncertainties, there is a saving grace that makes the design of offshore platforms safe and reliable. About 90% of the joints have fatigue life in excess of 1000 years. S–n curve is a conservative estimate which has a substantial reserve. Periodic underwater inspection, which is carried out on critical joints should be helpful in arresting crack propagation.

2.10 RELIABILITY PROBLEM

Reliability analysis of offshore structures could be formulated in different ways. Formulations can be broadly divided into two groups namely: (i) time-invariant problems and (ii) time-variant problems. In both, a limit state function is defined, which could be based on either serviceability or ultimate stress criteria. The reliability problem aims to determine the probability of limit state failure, which violates the (assumed) limit state condition. For the time-invariant problem, let $g(x) = 0$ describes the limit state function in which x denotes a set of random variables $x_1, x_2, \ldots, x_n$ and $g(x) \leq 0$ denotes the failure event. The probability of failure is then defined as

$$P_f = P[g(x) \leq 0] = \int_D f(x)dx \tag{2.4}$$

in which $f(x)$ is the joint probability density function of $g(x) = 0$ and the integration is performed over the domain D where $g(x) < 0$. The reliability is defined as

$$R = 1 - P_f \tag{2.5}$$

Figure 2.5 shows the concept for the special case of two random variables. Integration of a suitable probability density function within the shaded area is the reliability against failure, which is defined by the limit state function $(x_1, x_2) = 0$. A reliability problem is said to be time variant if the limit state function is also a function of time. In more specific terms, limit state function is defined by $g(x, y(t))$, in which $\{x\}$ is a set of random variables and $y(t)$ denotes a vector of stochastic processes. The failure event for such problems may constitute the out crossing of the vector processes, $y(t)$ through the limit state surface $g(x,y) = 0$ as shown in Figure 2.5. Probability of failure is given by

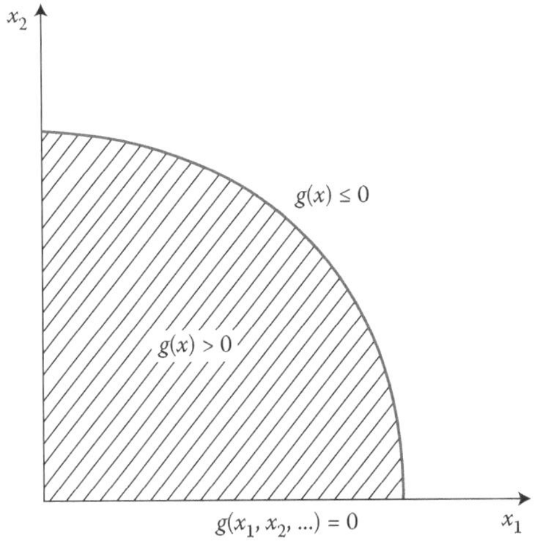

FIGURE 2.5 Concept of reliability with two random variables.

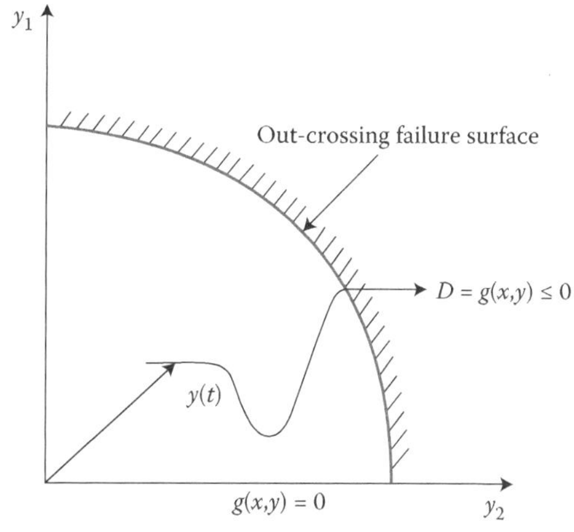

FIGURE 2.6 Time-variant reliability problem.

$$P_f = \int_{0<t\leq T} P[\min g(x, y(t) \leq 0) \mid x \mid f(x)dx] \tag{2.6}$$

where T denotes the life time of the structure, $f(x)$ is the joint probability density function of x. Conditional probability, which is defined by the terms within parenthesis is obtained by the stochastic analysis of structure for the random loading $y(t)$. Exact solution of the out crossing analysis of Equation 2.6 is difficult because of two reasons: (i) evaluation of the conditional probability; and (ii) determination of joint probability density function $f(x)$. In fact, the latter is also the reason for not being able to obtain the exact integration of Equation 2.6. Figure 2.6 shows the failure domain of a time-variant reliability problem.

2.11 RELIABILITY METHODS

There are many methods of finding the probabilities of failure of structures involving functions of random variables. Accuracies of the methods depend upon two critical issues: (i) how accurately the joint probability functions of the random variables are determined; and (ii) how accurately the integrations are evaluated. Furthermore, based on the methods to estimate the joint probability functions, they can also be classified as analytical or numerical. Following sections describe a few of the methods that are commonly used.

2.12 FIRST-ORDER SECOND MOMENT METHOD

In the first-order second moment method (FOSM), a first-order Taylor series approximation of the limit state function is used. Only second moment statistics of the random variables are employed to obtain the probability of failure. In its original form, two random variables are used to derive the method. The limit state function is defined as

$$Z = R - S \tag{2.7}$$

Assuming that R and S are statistically independent and normally distributed random variables, Z is also normally distributed. Its mean and covariance are obtained as

$$\begin{aligned} \mu_Z &= \mu_R - \mu_S \\ \sigma_Z^2 &= \sigma_R^2 + \sigma_S^2 \end{aligned} \tag{2.8}$$

The probability of failure is given by

$$P_f = P[Z < 0] = P[(R - S) < 0] \tag{2.9}$$

If Z is a normal variate, then it may be easily shown that P_f is given by

$$P_f = \Phi\left(-\frac{\mu_Z}{\sigma_Z}\right) \tag{2.10}$$

where Φ is the cumulative distribution function for a standard normal variable. Substituting for μ_Z and σ_Z, P_f can be rewritten as

$$P_f = 1 - \Phi\left(\frac{\mu_R - \mu_S}{\sqrt{\sigma_R^2 + \sigma_S^2}}\right) \tag{2.11}$$

The ratio of μ_Z/σ_Z is denoted by β and is well known as reliability index (safety index) in the theory of reliability. Then, P_f is also popularly expressed as

$$P_f = \Phi(-\beta) \tag{2.12}$$

If the variables R and S are log normally distributed, then the limit state function is defined as

$$Z = \ln\left(\frac{R}{S}\right) \tag{2.13}$$

where Z is a normal variable and the probability of failure can be expressed as

$$P_f = 1 - \Phi\left\{\frac{\ln(\mu_R/\mu_S)\sqrt{(1+\delta_S^2)/(1+\delta_R^2)}}{\sqrt{\ln(1+\delta_R^2)\ln(1+\delta_S^2)}}\right\} \tag{2.14}$$

where δ_R and δ_S are the coefficient variations of R and S. The above formulation may be generalized to many random variables, denoted by a vector **X**. Let the performance function be written as

$$Z = G(X) \tag{2.15}$$

The Taylor series expansion of the performance function about the mean values is given by

$$Z = G(\bar{X}) + \sum_{i=1}^{N} \frac{\partial G}{\partial x_i}(x_i - \bar{x}_i) + \frac{1}{2}\sum_{i=1}^{N}\sum_{j=1}^{N} \frac{\partial^2 G}{\partial x_i\, \partial x_j}(x_i - \bar{x}_i)(x_j - \bar{x}_j) \tag{2.16}$$

where $\bar{x}_i$ is the mean of the variable x_i. Truncating the series at the linear terms, the first-order approximation of mean and variance of Z are obtained as

$$\bar{Z} = G(\bar{X}) \tag{2.17}$$

$$\sigma_Z^2 = \sum_{i=1}^{n}\sum_{j=1}^{n} \frac{\partial^2 G}{\partial x_i\, \partial x_j} \operatorname{cov}(x_i x_j) \tag{2.18}$$

where $\operatorname{cov}(x_i, x_j)$ is the covariance of x_i and x_j. If the variables are assumed to be statistically independent, then, the following relationship holds good:

$$\sigma_Z^2 = \sum_{i=1}^{n} \left(\frac{\partial G}{\partial x_i}\right)^2 \sigma_{x_i}^2 \tag{2.19}$$

where partial derivatives of the above equation are obtained at the mean values.

2.13 HASOFER–LIND METHOD

While the method of finding first-order approximate mean and variance using the Taylor series expansion is valid for nonlinear performance function, the safety index thus obtained cannot be directly related to the exact probability of failure (Srinivasan Chandrasekaran, 2015a,b,c,d). The estimate of safety index will be more accurate when all random variables are statistically independent. Given the fact that normal variables and performance function are a linear combination of the random variables, the Hasofer–Lind method gives an improved safety index in comparison to the FOSM method. This method is centered on the computation of the design point. The design point is the minimum distance of performance function from the origin, which is the safety index for the case of normal random variables with linear performance function. The method uses reduced variables, which is defined as

$$x_i' = \frac{x_i - \bar{x}_i}{\sigma_{x_i}} \quad (i = 1, \ldots, n) \tag{2.20}$$

Thus, the reduced variable has zero mean and unit standard direction. With the help of the reduced variable, original limit state or performance function $G(X) = 0$ is converted to $(X') = 0$. The minimum distance, called β_{HL}, can be expressed as

$$\beta_{HL} = [(X'_d)(X'_d)^T]^{1/2} \tag{2.21}$$

where X'_d is the minimum distance point on the limit state function and is called the design point or checking point.

The importance of finding β_{HL} is explained with the help of the linear limit state function of two variables. Consider the limit state function as follows:

$$Z = R - S = 0 \tag{2.22}$$

The reduced variables are then defined as

$$R' = \frac{R - \mu_R}{\sigma_R} \tag{2.23}$$

$$S' = \frac{S - \mu_s}{\sigma_s} \tag{2.24}$$

Substituting for R and S, the limit state equation may be expressed in terms of R' and S' as

$$\sigma_R R' - \sigma_s S' + \mu_R - \mu_s \tag{2.25}$$

In the space of reduced variable, limit state function can be plotted as shown in Figure 2.7. It is apparent from the figure that if the limit state is linear near the origin,

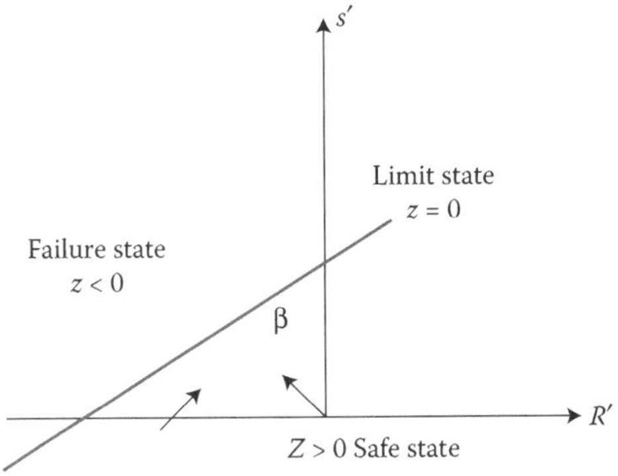

FIGURE 2.7 Hasofer–Lind reliability index with linear performance function.

the failure region is larger and the probability of failure is more. The minimum distance of the line from the origin is computed as

$$\beta_{HL} = \frac{\mu_R - \mu_S}{\sqrt{\sigma_R^2 + \sigma_S^2}} \tag{2.26}$$

Note that β_{HL} is the same as β defined for normal variables R and S. Thus, β_{HL} can be regarded as a measure of the safety index.

The Hasofer–Lind reliability index can be used to calculate a first-order approximation to the failure probability as $P_f = \Phi(-\beta_{HL})$. This is the integration of the standard normal density along the ray joining the origin and X'_d. When the limit state function is nonlinear, the computation of the minimum distance becomes an optimization problem:

$$\text{Minimize } D = \sqrt{(X')^T(X')} \tag{2.27}$$

Subject to the following condition:

$$G(X') = 0 \tag{2.28}$$

Shinozuka obtained an expression for the minimum distance as given below

$$\beta_{HL} = -\frac{\sum_{i=1}^{n} x'_{di}(\partial G/\partial x'_{di})}{\sqrt{\sum_{i=1}^{n} (\partial G/\partial x'_{di})^2}} \tag{2.29}$$

where the derivatives $(\partial G/\partial x'_{di})$ are evaluated at the design point $x'_{d1}, x'_{d2}, \ldots, x'_{dn}$. The design point is given by

$$x'_{di} = \alpha_{di}\beta_{HL}; \quad (i = 1, 2, \ldots, n) \tag{2.30}$$

where

$$\alpha_{di} = \frac{(\partial G/\partial x'_{di})}{\sqrt{\sum_{i=1}^{n} (\partial G/\partial x'_{di})^2}}$$

are the direction cosines along the coordinate axes x'_i. In the space of original variables, the design point is given by

$$x'_{di} = \mu_{x_i} - \alpha_{di}\sigma_{x_i}\beta_{HL} \tag{2.31}$$

The method results in a linear approximation of limit state function at the design point as shown in Figure 2.8. With the β_{HL} as obtained above, an estimate of P_f is given by $P_f = \Phi(-\beta_{HL})$. If all the variables are not normally distributed, then it is difficult to relate β_{HL} to the exact probability of failure. Rackwitz and Fiessler suggested that an improved estimate of the probability of failure may be obtained by transforming the non-normal variables into equivalent normal variables. They estimated the parameters of the equivalent normal distribution $\mu_{x_i}^N$ and $\sigma_{x_i}^N$ by imposing two conditions namely: (i) the cumulative distribution functions, and (ii) the probability density functions of the original variables and the equivalent normal variables should be equal at the checking point $(x'_{d1}, x'_{d2}, \ldots, x'_{dn})$ on the limit state surface. The mean values and the standard deviations of the equivalent normal variables are given by

$$\sigma_{x_i}^N = \frac{\bar{\Phi}\{\Phi^{-1}[F_i(x_{di})]\}}{f_i(x_{di})} \tag{2.32}$$

$$\mu_{x_i}^N = x_{di} - \Phi^{-1}[F_i(x_{di})]\sigma_{x_i}^N \tag{2.33}$$

where F_i and $\bar{F}_i$ are the non-normal cumulative distribution and density functions of x_i; and Φ and $\bar{\Phi}$ are the cumulative distribution and density functions of the standard normal variate, respectively.

With this modification of the random variables, the following algorithm may be used to compute β_{HL}:

i. Assume initial values of the design point x_{di} ($i = 1,2,\ldots,n$). To start with, the mean values of the random variables may be considered as the design point.

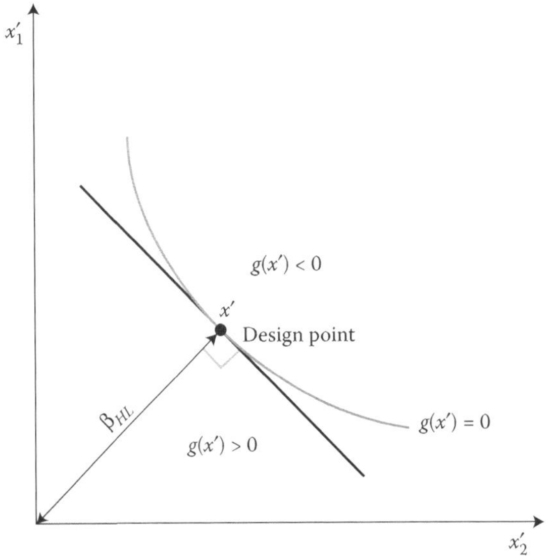

FIGURE 2.8 Hasofer–Lind reliability index with nonlinear performance function.

ii. Obtain the reduced variable $x'_{di} = (x_{di} - \mu_{x_i})/\sigma_{x_i}$.
iii. Evaluate $(\partial G/\partial x'_{di})$ and α_{di} at x'_{di}.
iv. Obtain the new design point x_{di} in terms of β_{HL} (use Equation 2.26).
v. Substitute the new x_{di} in the limit state equation $G(X_d) = 0$ and solve for β_{HL}.
vi. Using the β_{HL} value obtained in Step v, evaluate $x'_{di} = -\alpha_{di}\beta_{HL}$.
vii. Repeat Steps iii through vi till convergence is achieved.
viii. Obtain $\sigma_{x_i}^N$ and $\mu_{X_i}^N$ (use Equations 2.28 and 2.29).
ix. Use Equation 2.27 to obtain equivalent normal variables in terms of β_{HL}. Note that μ_{xi} and σ_{xi} in Equation 2.27 should be replaced by $\mu_{X_i}^N$ and $\sigma_{x_i}^N$.
x. Solve $G(X_d) = 0$ to obtain β_{HL}.
xi. Obtain $p_f = \Phi(-\beta_{HL})$.

2.14 SECOND-ORDER RELIABILITY METHODS

The limit state function could be nonlinear because of many reasons such as (i) nonlinear relationship between random variables and the limit state function; (ii) transformation of non-normal variables to standard normal variables; and (iii) transformation from correlated to uncorrelated variables. In such cases, the joint probability density function does not decay rapidly as it moves away from the minimum distance point. Higher order approximation is required for failure probability computations. In this method, the curvature of the limit state function around the minimum distance point is approximated by a quadratic function. Fiessler et al. explored the use of various quadratic approximations. A closed-form solution for the probability of failure of a region, bounded by a quadratic limit state is given by Breitung using asymptotic approximations as given below

$$p_f \approx \Phi(-\beta)\prod_{i=1}^{n-1} 1 + \beta K_i^{-1/2} \tag{2.34}$$

where k_i denotes the ith main curvature of the limit state function at the minimum distance point. The main curvature k_i is the ith eigen value of the second derivative matrix **A** of the limit state surface at the design point, in a rotated normal space. The elements of the matrix **A** is given by

$$a_{ij} = \frac{(RDR^T)}{|\nabla G(Y^*)|} \quad (i, j = 1, 2, \ldots, n-1) \tag{2.35}$$

where R is the orthogonal transformation matrix for which nth row is selected to be $Y^*/(Y^{*T}Y^*)^{1/2}$. A standard Gram–Schmidt algorithm may be used to determine R; D is the $n \times n$ second derivative matrix of the limit state surface in the standard normal space evaluated at the design point; $\nabla G(Y^*)$ is the gradient vector in the standard normal space.

2.15 SIMULATION-BASED RELIABILITY METHOD

Let the performance function or limit state function be given by $g(X) = 0$. The probability of failure P_f is obtained as

$$P_f = \iint \dots \int_{g(x)\le 0} f_x(X)dx \tag{2.36}$$

where $f_x(X)$ is the joint density function of variables $x_1, x_2, \dots, x_n$ and $d\mathbf{X}$ stands for $dx_1, dx_2, dx_3, \dots, dx_n$. The simulation procedure consists of the Monte-Carlo simulation of basic variables according to their probabilistic characteristics. The number of failures N_f (i.e., $g(X) < 0$) is counted for the set of random variables generated. The probability of failure is given by

$$P_f = \frac{N_f}{N} \tag{2.37}$$

where N is the total number of simulation cycles. The estimated probability of failure depends upon the number of cycles of simulation used. For a sufficiently accurate result, a large number of simulations may be needed. Therefore, it is better to approximately compute the variance of the estimated probability of failure. This is done by assuming each simulation cycle to constitute a Bernoulli trial. Therefore, N_f in N trials can be considered to follow a binomial distribution. Variance of the estimated probability of failure can be computed approximately as

$$\text{var}(P_f) = \frac{(1-P_f)}{N} \tag{2.38}$$

The statistical accuracy of the estimated P_f is measured by the coefficient of variation given by

$$\text{cov}(p_f) = \frac{\sqrt{((1-p_f)p_f)/N}}{P_f} \tag{2.39}$$

It is important to note that smaller the coefficient of variation the better is the accuracy; accordingly, N is decided.

2.16 RELIABILITY ESTIMATE USING HIGHER-ORDER RESPONSE SURFACE METHODS

Structural reliability analysis is performed to determine safety of the system (design) under various loading states, material characteristics, and structural forms. However, for large and complex systems like offshore structures, approximate techniques are also used in order to reduce computational efforts to an acceptable level.

The response surface method is a recent development, which can provide a reasonable estimate of reliability regardless of the complexity of the failure process of the structural system (Tvedt, 1990). This method is a mixture of statistical and mathematical techniques that are useful in modeling and analyzing the structure whose response is influenced by several variables; the objective is to optimize this response under the given conditions. The method approximates the limit state functions by simple and explicit techniques and thereby avoids the true input–output relationship during simulations. Limit state function $g(X)$, which is an implicit function of the basic random variable X is replaced by nth order polynomial function $\hat{g}(X)$ with undetermined coefficients. The selection of the order of polynomial should be done carefully such that degree of $\hat{g}(X)$ should be lesser than or equal to $g(X)$, which will lead to a well-conditioned system of linear equations with unknown coefficients. In reliability analysis, it is quite common that neither limit state function nor design points are known, which will lead to a confusing state of choosing the degree of polynomial; typically a quadratic, with or without cross-terms, is used for approximation and is given by

$$\hat{g}(X) = a_0 + \sum_{i=1}^{m} a_i x_i + \sum_{i=1}^{m} a_{ii} x_i^2 + \sum_{i=1}^{m}\sum_{j=1}^{m} a_{ij} x_i x_j \quad \text{for } (i < j) \tag{2.40}$$

where m is the number of random variables X and a_0, a_i, a_{ii}, a_{ij} are unknown coefficients, which are obtained from discrete evaluations of implicit limit state function through finite element routine. The resulting fitted surface is the explicit equivalent of the implicit limit state function, on which reliability methods can be applied. More importantly, the selection of design points, where this function will be evaluated is crucial. Bucher and Bourgund (1990) proposed an iterative response surface approach, where successive iteration shifts the design points toward the limit state function and is given by

$$x_m = \bar{\mu} + (x_D - \bar{\mu}) \frac{g(\bar{\mu})}{[g(\bar{\mu}) - g(x_D)]} \tag{2.41}$$

where $\bar{\mu}$, x_D are mean vector and minimum norm points for the limit state $\hat{g}(X) = 0$

2.17 HIGH-ORDER STOCHASTIC RESPONSE SURFACE METHOD

A quadratic polynomial may not be enough to describe higher-order limit states. In such cases, it is necessary to perform statistical analyses of trial response surfaces to determine the appropriate order of approximation, which is given by

$$\tilde{g}(X) = a + \sum_{i=1}^{n}\sum_{j=1}^{k_i} b_{ij} X_i^j + \sum_{q=1}^{m} c_q \prod_{i=1}^{n} X_i^{p_{iq}} \tag{2.42}$$

where coefficients b_{ij} correspond to the terms involving only one random variable and coefficients c_q correspond to mixed terms, involving the product of two or more random variables. The polynomial order, k_i, total number of mixed terms, m, and order of random variable in a mixed term, p_{iq} are determined in the various stages. The algorithm of the method consists of four stages: (i) identifying order of response surface; (ii) determining number and type of mixed terms; (iii) estimating coefficients of high order of response surface polynomial; and (iv) determining probability of failure with response surface (Cronin et al., 1978; Crow and Shimizu, 1988). Monte-Carlo simulation is carried out on the response surface to determine the probability of failure.

In the first stage, polynomial order k_i is determined by numerically testing the significance of polynomial coefficients along the coordinate axes x_i. The Chebyshev polynomial is selected as the basis function, which is orthogonal and bounded within the range of [−1,1]. A Chebyshev polynomial of degree M is given by

$$T_M(x) = \cos(M \arccos(x)) \tag{2.43}$$

where $\min(T_M(x)) = -1$ and $\max(T_M(x)) = 1$ for all values of variable such that $-1 \leq x \leq 1$. Polynomial has M roots in the interval [−1, 1] and are given by

$$x_m = \cos\left\{\frac{\pi(m-(1/2))}{M}\right\} \quad \text{where } m = 1,2,3,\ldots,M \tag{2.44}$$

Orders of variables k_i of Equation 2.42 are estimated one-by-one along the dimension X_i using the one-dimensional Chebyshev polynomial, which is given by

$$\hat{g}(X) = d_0T_0(x_i) + d_1T_1(x_i) + d_2T_2(x_i) + \cdots + d_nT_n(x_i) \tag{2.45}$$

where $T_{jk} = T_j(x_k)$, x_k is the kth root of $T_k(x)$, x_i is interpolated values of X_i from interval $[\mu - h_{ord}\,\sigma_i,\ \mu + h_{ord}\,\sigma_i]$ to [−1,1]. h_{ord} is the domain of sampling points used to determine the polynomial degree of the approximation. Chebyshev coefficients d_j are determined by the least-squares method:

$$D = [T^TT]^{-1}T^Tg_i(x_i) \tag{2.46}$$

Since all Chebyshev polynomial coefficients are bounded within [−1,1], contribution of T_j to the limit state function is related to the order of d_j only.

2.18 SYSTEM RELIABILITY

So far, we have considered structural component reliability in which the performance of a structural element is quantified as a single, continuous, and differentiable limit state function (Elishakoff, 1999). In case more than one function is needed to define the performance of a structure or that function is not continuous and differentiable, then the structure must be analyzed as a system. Hence, a system is an assemblage of components, each having a limit state function associated with it that defines the performance of the respective component. Similar to the component reliability problems, which considered only two states (safety and failure), systems will also have only two states. Consider a system consisting of *n* components. For the *i*th component, the variable is defined as

$$a_i = \begin{Bmatrix} 1 \\ 0 \end{Bmatrix} \tag{2.47}$$

The above expression is valid if *i*th component does not perform the intended function and has failed. Similarly, for the system, it is defined as

$$a_z = \begin{cases} 1 \\ 0 \end{cases} \tag{2.48}$$

where 1 represents if the system is functional and 0 if the system fails. The random variable a_i is an indicator function that specifies the performance state of the *i*th component. It should be apparent that the probability that $a_1 = 0$ can be estimated by a component reliability analysis. The indicator function *a*, which specifies the performance state of the system is a function of the performance of its components and is given by

$$a_Z = \Psi(a) \tag{2.49}$$

where $a = [a_1, a_2, a_3, \ldots, a_n]^T$ $\Psi()$ denotes the system auction. While the performance criterion of an individual element is usually easy to define and quantify through a limit state function, performance criteria for a system is complex. An important step in system reliability analyses is the identification of all combinations of component failures that constitute a failure of the system. Systems are idealized in the following ways.

2.18.1 Series System

A system that fails if any of its components fails is called a series system. Such systems are often called weakest link systems. For example, a chain is considered to be as strong as the weakest link. Function for a series system is given by

$$a_z = \Psi(a) = \min(a_1, a_2, \ldots, a_n) = \prod_{i=1}^{n} a_i \tag{2.50}$$

Clearly, if any component fails, then the product term in Equation 2.50 is equal to zero.

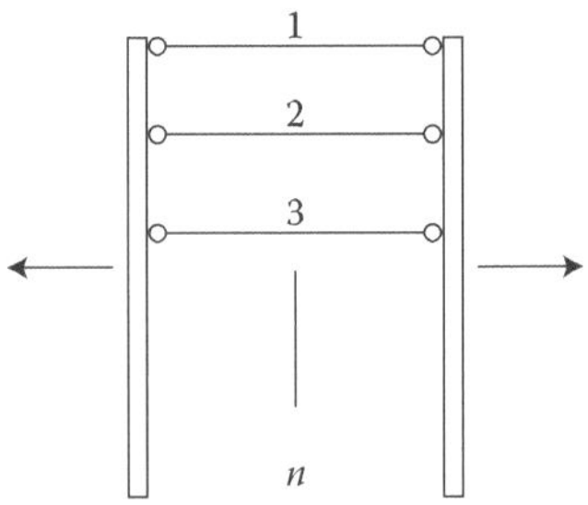

FIGURE 2.9 Parallel system consisting of n components.

2.18.2 Parallel System

A system that fails only if all of its components fail is called a parallel system. Such systems are represented schematically as shown in Figure 2.9.

Function for a parallel system is given by

$$a_z = \Psi(a) = \max(a_1, a_2, \ldots, a_n) = 1 - \prod_{i=1}^{n}(1 - a_j) \tag{2.51}$$

It is seen that Equation 2.51 is zero only when $a_i = 0$ for $i = 1,2, \ldots, n$ (when all components have failed). If even one component functions, then the product term in Equation 2.51 will be equal to zero and the system function is unity.

2.18.3 k-out-of-n System

When k of n components of a system functions, then it is called a k-out-of-n system. Function of the system, in this case is given by

$$a_i = \begin{Bmatrix} 1 & \text{if} \sum_{i=1}^{n} a_1 \geq K \\ 0 & \end{Bmatrix} \tag{2.52}$$

2.19 GENERAL SYSTEMS

General systems are combination of the above systems, as shown in Figure 2.10. A general system is said to have failed if a force (or deformation, etc.) can be transmitted from one end of the system to the other. To facilitate the analysis of general systems, it is useful to introduce the concepts of cut sets and path sets.

2.19.1 Cut Sets

A cut set is any set of components whose joint failure constitutes a failure of the system. For the general system shown above, cut sets are:

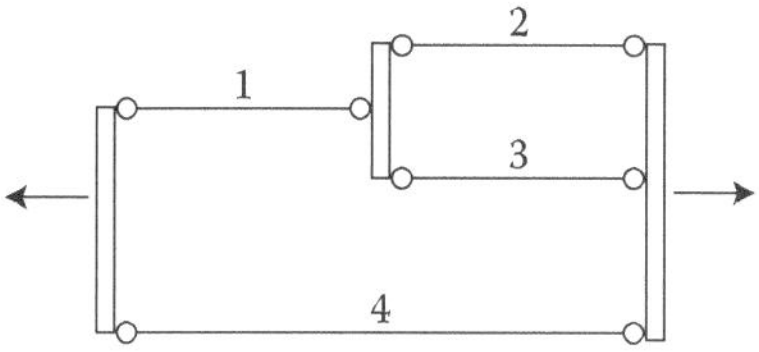

FIGURE 2.10 A general system.

$$C_1 = \{1,2,3,4\}\ C_2 = \{1,2,4\}\ C_3 = \{1,3,4\}\ C_4 = \{2,3,4\}\ C_5 = \{1,4\}$$

A *minimum cut set* is a cut set that upon removal of each of its components (one at a time) ceases to be a cut set. Thus, for the above example, the minimum cut sets are C_4 and C_5. It is interesting to note that C_1 is not a minimum cut set because we can remove component 1 and still have a cut set (C_4). Similarly, removing component 2 from cut set C_2 or component 3 from cut set C_3 do not cause either of these sets to cease to be a cut set. *Disjoint cut sets* are cut sets that do not contain common components. For the above example, there are no disjoint cut sets because each cut set contains component 4.

2.19.2 Path Sets

An alternative to the cut set formulation of a general system is the path set. A path set is any set of components whose joint survival constitutes survival of the system. For the general system considered above, the path sets are

$$P_1 = \{4\}\ P_2 = \{1,2\}\ P_3 = \{1,3\}\ P_4 = \{1,4\}\ P_5 = \{2,4\}$$

$$P_6 = \{3,4\}\ P_7 = \{1,2,3\}\ P_8 = \{1,2,4\}\ P_9 = \{1,3,4\}$$

$$P_{10} = \{2,3,4\}\ P_{11} = \{1,2,3,4\}$$

A minimum path set is a path set that upon the removal of each of its components (one at a time) ceases to be a path set. Therefore, for the above example, the minimum path sets are P_1, P_2, and P_3. Note that for the remaining path sets, there is at least one component that can be removed without causing the set to cease to be a path set. *Disjoint path sets* are path sets that do not contain common components. For the above example, P_1 and P_2 are disjoint path sets.

2.20 SYSTEM FUNCTIONS FOR GENERAL SYSTEMS

General systems are assumed to be a combination of either a series system with its minimum cut sets or a parallel system with its minimum path sets. These representations of the general system are shown in Figures 2.11 and 2.12, respectively.

As shown above, each minimum cut set can be thought of as a subsystem of parallel components and each minimum path set can be thought of as a subsystem of components

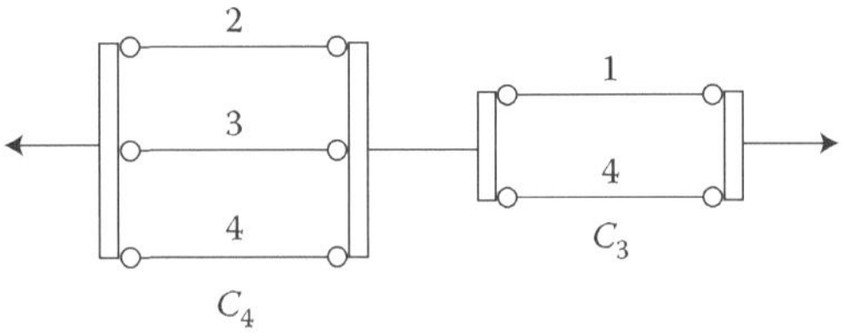

FIGURE 2.11 General system with minimum cut sets.

in series. For the cut set formulation, the system fails if components 2, 3, and 4 fail simultaneously and/or components 1 and 4 fail simultaneously. No other combinations of component failures constitute a failure of the system. For the path set formulation, it is easy to verify that the system survives if any one of the series subsystems shown in Figure 2.12 survives. Function for the general system considered above is given by

$$a_Z = \prod_{m-1}^{n_c} \alpha_m(a) \tag{2.53}$$

where n_c is the number of cut set and $\alpha_m(a)$ is the cut set function of the mth cut set. Note that Equation 2.53 has a form similar to that of Equation 2.50 for a series system. The cut set function is given by

$$\alpha_m(a) = 1 - \prod_{i=c_m}(1 - a_i) \tag{2.54}$$

where C_m represents the mth cut set. Thus, for the general system, the system function is given by

$$a_z = [1-(1-a_2)(1-a_3)(1-a_4)][(1-(1-a_1))(1-(a_4))] \tag{2.55}$$

For Boolean variables, the following property is valid. Using this and expanding it, we get

$$a_i^K = a_i, \quad K > 0 \tag{2.56}$$

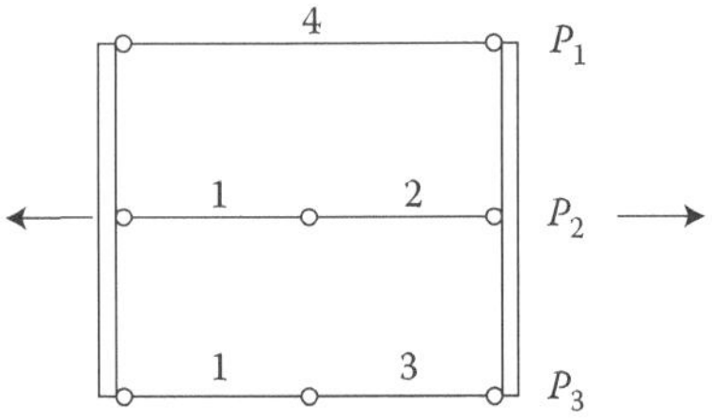

FIGURE 2.12 General system with minimum path sets.

$$a_z = a_4 + a_1a_2 + a_1a_3 - a_1a_2a_3 - a_1a_2a_4 - a_1a_2a_3 + a_1a_2a_3a_4 \tag{2.57}$$

Similarly, for the general system with minimum path sets, the system function is given by

$$a_z^{1} = 1 - \prod_{i=1}^{n_p} [1 - \beta_1(a)] \tag{2.58}$$

where n_p is the number of path set and β_1 (a) is the path set function of the ith path set. The path set function is given by

$$\beta_1(a) = \prod_{i \in p_i} a_i \tag{2.59}$$

where P_i is the ith path set. Thus, for the general system, the function is given by

$$a_Z = 1 - [1 - a_4] | 1 - a_1a_2 | 1 - a_1a_3 \tag{2.60}$$

Expanding, we get

$$a_z = a_4 + a_1a_2 + a_1a_3 - a_1a_2a_3 - a_1a_2a_3 - a_1a_2a_4 - a_1a_3a_4 + a_1a_2a_3a_4 \tag{2.61}$$

2.21 COMPUTING SYSTEM RELIABILITY

For a_{ij}, $i = 1,2, \ldots, n$ for the components and a_z for the system are Bernoulli variables, let $P_i = P(a_j = 0)$ denote the probability of failure of ith component and $PF = P(a_z = 0)$ denote the probability of failure of the system. For the ith component, the following relationship holds good:

$$E[a_i] = 1(1 - p_i) + 0(p_i) = 1 - p_i \tag{2.62}$$

Hence,

$$P_i = E[\bar{a}_j] \tag{2.63}$$

where $\bar{a}_i = 1 - a_i$ is the complementary event to a_i. Similarly,

$$P_F = E[\bar{a}_z] = 1 - E[a_Z] = 1 - E[\Psi(a)] \tag{2.64}$$

where $\bar{a}_Z = 1 - a_Z$

In terms of component reliability analyses, event $\{a_i = 0\}$ is equivalent to $\{g_i(x) \le 0\}$. In other words, ith component fails when the outcomes of the random variables in a given problem define a point in the failure domain of the component. For a series system consisting of n components, the system fails if any of its components fail. Thus, the failure domain of a series system, Ω_z, is the union of the component failure domains and is given by

$$\Omega_Z = \left\{ \bigcup_{i=1}^{n} (g)X \le 0 \right\} \tag{2.65}$$

For a parallel system consisting of n components, the system fails if all of its components fail. Failure domain for a parallel system Ω_p is the intersection of the component failure domains and is given by

$$\Omega_p = \bigcap_{i=1}^{n} \{g_i(x) \le 0\} \tag{2.66}$$

Figure 2.13 shows the failure domain for the series system and Figure 2.14 shows that of the parallel system consisting of three components. In the figure, hatching

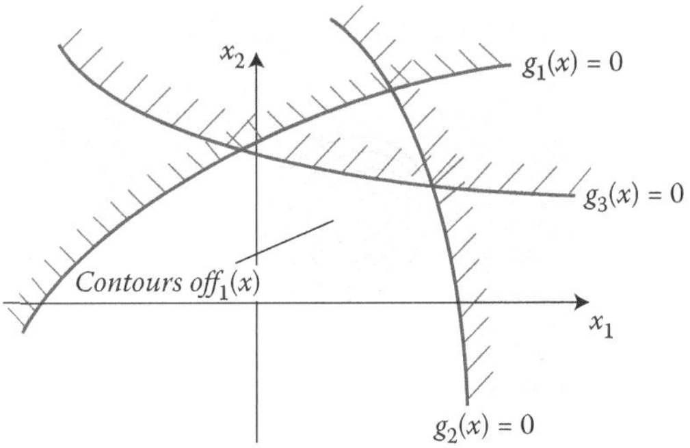

FIGURE 2.13 Failure domains for a series system consisting of three components.

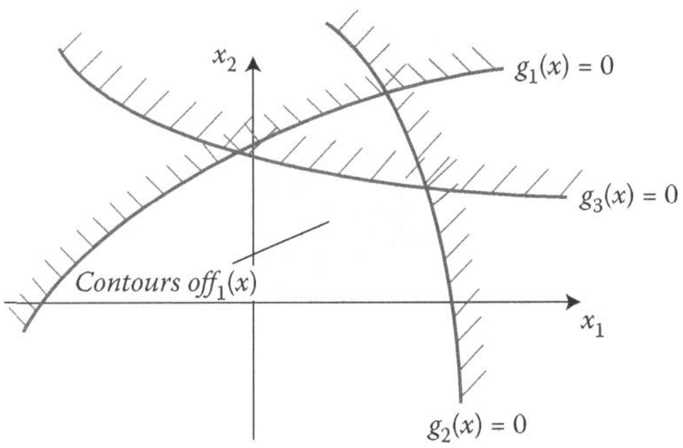

FIGURE 2.14 Failure domains for a parallel system consisting of three components.

along the limit state surfaces indicates the region in which the corresponding limit state function is less than zero.

For a general system, the failure domain, Ω_z is given by the cut set formulation as

$$\Omega_g = \bigcup_m \bigcap_{i \in c_m} \{g_i(x) \leq 0\} \tag{2.67}$$

where C_m is the mth cut set. Safe domain for a general system $\bar{\Omega}_g$ is defined by the path set formulation and given by

$$\bar{\Omega}_g = \bigcup_i \bigcap_{i \in p_i} \{g_i(x) > 0\} \tag{2.68}$$

where P_i is the ith path set. For the general system, failure domain is given by the union of the minimum cut sets as

$$C_4 = \{2,3,4\} \text{ and } C_5 = \{1,4\}$$

This failure domain is shaded in Figure 2.15 for some assumed component limit state surfaces. The problem of system reliability can be stated as

$$P_F = \int_{\Omega_g} f(x\,dx) \tag{2.69}$$

Alternatively,

$$1 - P_F = \int_{\Omega_R} f(x)dx \tag{2.70}$$

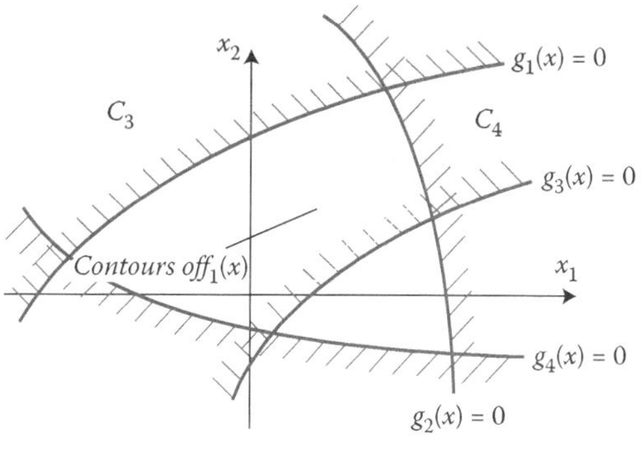

FIGURE 2.15 Failure domain for the general system.

2.22 FIRST-ORDER ESTIMATES

In order to compute system reliability, one must be able to compute the probability of a union of events (series systems or subsystems) or the probability of an intersection of events (parallel systems or subsystems). For a series system, the probability of failure is given by

$$P_F = P\left(\bigcup_{i=1}^{n}\{g_i(x) \le 0\}\right) \tag{2.71}$$

After applying the appropriate transformation $u = u(x)$ to the standard normal space, we get

$$P_F \approx P\left(\bigcup_{i=1}^{n}\{G_i(u) \le 0\}\right) \tag{2.72}$$

where an approximation due to the mapping of non-normal variables to the standard normal space is introduced. Linearizing, $G_i(u) = 0$ at the design point for the ith limit state function using Taylor series expansion, we get

$$\begin{aligned} G_i(u) &\approx \nabla G^T\left(u_i^*\right)\left(u-u_i^*\right) = \left\|\nabla G_i\left(u_i^*\right)\right\|\left[-\alpha_i^T\left(u-u_i^*\right)\right] \\ &= \left\|\nabla G_i\left(u_i^*\right)\right\|\left[\beta_i - \alpha_i^T u\right] \end{aligned} \tag{2.73}$$

where u_i^* and β_i are the design point and reliability index, respectively for the ith component (obtained from a FORM analysis of the component); α_i is the corresponding unit normal vector to the limit state surface $G_i(u) = 0$ at u_i^*. Substituting Equation 2.73 into Equation 2.72, we get

$$P_F \approx P\left[\bigcup_{i=1}^{n}\left\{\left(\beta_i - \alpha_{iu}^T\right) \le 0\right\}\right] \tag{2.74}$$

After dividing both sides of the inequality by the positive scalar $\| \nabla G_i(u_i^*) \|$ we now define

$$Z_i = -\alpha_i^T \sim N(0,1) \tag{2.75}$$

which is a standard normal variable. Hence, probability of failure can be rewritten as

$$P_F \approx P\left(\bigcup_{i=1}^{n}\{Z_i \le -\beta_i\}\right) \tag{2.76}$$

Using de Morgan's laws and making use of the rotational symmetry of the standard normal space, we get

$$\begin{aligned}P_F &\approx P\left(\bigcup_{i=1}^{n}\{Z_i \le -\beta_i\}\right) = 1 - P\left(\bigcap_{i=1}^{n}\{Z_i > -\beta_i\}\right)\\ &= 1 - P\left(\bigcap_{i=1}^{n}\{Z_i \le -\beta_i\}\right)\\ &= 1 - \Phi_n(\beta, R_{ZZ})\end{aligned} \tag{2.77}$$

where $\Phi_n(\beta, R_{ZZ})$ is the joint normal cumulative density function (cdf) with correlation matrix R_{ZZ} evaluated at $\beta = [\beta_1, \beta_2, \ldots, \beta_n]^T$. In the present case, the correlation matrix is identical to the covariance matrix as the variables are standard normal. The correlation matrix is given by

$$R_{ZZ} = \Sigma_{ZZ} = A\Sigma_{UU}A^T = AA^T \tag{2.78}$$

where the ith row of A is α_i^T and $\Sigma_{UU} = 1$ due to the definition of standard normal variables. Thus, the off-diagonal terms in correlation matrix R_{ZZ} are given by

$$\rho Z_i Z_j = \alpha_i^T \alpha_j \tag{2.79}$$

which quantifies the correlation between failure modes i and j. Kindly note that unlike in component reliability analyses, where the unit normal vector α is of secondary importance due to the rotational symmetry of the standard normal space, in system analyses, the relative directions of the unit normal α_i and α_j play a significant role.

For a parallel system, the probability of failure is given by

$$P_F = P\left(\bigcap_{i=1}^{n}\{g_i(x) \le 0\}\right) \tag{2.80}$$

This is approximated as

$$\begin{aligned}P_F &= P\left(\bigcap_{i=1}^{n}\{g_i(x) \le 0\}\right) \approx P\left(\bigcap_{i=1}^{n}\{G_i(u) \le 0\}\right)\\ &\approx P\left(\bigcap_{i=1}^{n}\left\{\left(\beta_i - \alpha_i^T u\right) \le 0\right\}\right)\\ &\approx P\left(\bigcap_{i=1}^{n}\left\{Z_i \le -\beta_i\right\}\right)\\ &\approx \Phi_n(-\beta, R_{ZZ})\end{aligned} \tag{2.81}$$

in which $\Phi_n(\cdot)$, β, and R_{ZZ} are as defined earlier.

The behavior of a general system can be modeled as a parallel system composed of path sets, with each path set acting like a subsystem of components in series. The first-order approximation for a general system's reliability, based on the cut set formulation, is presented below. A similar approach can be developed using the path set formulation. In order to estimate the probability of failure of a general system using the cut set formulation, one need to evaluate the probability of failure, which is given by

$$P_F = P\left(\bigcup_m \bigcap_{i \in c_m} \{g_i(x) \leq 0\}\right) \tag{2.82}$$

Let the event $E_m = \cap_{i \in c_m} \{g_i(x) \leq 0\}$ that is, E_m is the event that the parallel subsystem represented by cut set C_m fails. Hence, probability of failure can be rewritten as

$$P_F = P\left(\bigcup_m E_m\right) \tag{2.83}$$

As per the inclusion–exclusion of set theory, the following statements are derived:

$$\begin{aligned} P_F &= P\left(\bigcup_m E_1\right) = P\left(E_1 \cup E_2 \cup \ldots \cup E_{n_e}\right) \\ &= \sum_{i=1}^{n_e} P(E_i) - \sum_{j=1}^{n_e} \sum_{i=1}^{j-1} P(E_i E_j) \end{aligned} \tag{2.84}$$

where n_c is the number of minimum cut sets identified for the system. Thus, the above equation can be solved by summing the probabilities of failure of each cut set, $\sum_{i=1}^{n_e} P(E_i)$ along with the probabilities of failure of every possible intersection of cut sets (with the appropriate sign applied). For the general system used as an example earlier, two cut sets are $C_4 = \{2,3,4\}$ and $C_5 = \{1,4\}$. Let E_4, E_5 represent the events associated with the parallel subsystem where cut set C_4 and C_5 fails, respectively. The following relationship holds good:

$$E_4 = \{(g_2(X) \leq 0) \cap (g_3(x) \leq 0) \cap (g_4(x) \leq 0)\}$$

$$E_5 = \{(g_1(X)) \leq 0 \cap (g_4(x) \leq 0)\}$$

The probability of failure is given by

$$P_F = P(E_4) + P(E_5) - P(E_4 E_5) \tag{2.85}$$

where

$$P(E_4) \approx \varphi_3(-\beta_4, R_4) \quad (2.86)$$

$$P(E_5) \approx \phi_2(-\beta_5 R_5) \quad (2.87)$$

and $(E_4E_5) \approx \phi_4(-\beta_{45}\beta_{45})$
where

$$\beta_4 = \begin{Bmatrix} \beta_2 \\ \cdot \\ \beta_3 \\ \beta_4 \end{Bmatrix} R_4 = \begin{bmatrix} 1 & \alpha_2^T\alpha_3 & \alpha_2^T\alpha_4 \\ \alpha_3^T\alpha_2 & 1 & \alpha_3^T\alpha_4 \\ \alpha_4^T\alpha_1 & \alpha_4^T\alpha_3 & 1 \end{bmatrix}$$

$$\beta_5 = \begin{Bmatrix} \beta_1 \\ \beta_4 \end{Bmatrix} R_5 = \begin{bmatrix} 1 & \alpha_1^T\alpha_4 \\ \alpha_4^T\alpha_1 & 1 \end{bmatrix}$$

$$\beta_{45} = \begin{Bmatrix} \beta_1 \\ \beta_2 \\ \beta_3 \\ \beta_4 \end{Bmatrix} R_{45} = \begin{bmatrix} 1 & \alpha_1^T\alpha_2 & \alpha_1^T\alpha_3 & \alpha_1^T\alpha_4 \\ \alpha_2^T\alpha_1 & 1 & \alpha_2^T\alpha_3 & \alpha_2^T\alpha_4 \\ \alpha_3^T\alpha_1 & \alpha_3^T\alpha_2 & 1 & \alpha_3^T\alpha_4 \\ \alpha_4^T\alpha_1 & \alpha_4^T\alpha_2 & \alpha_4^T\alpha_3 & 1 \end{bmatrix}$$

Note that

$$E_4E_5 = E_4 \cap E_5 = \{(g_1(x) \le 0) \cap (g_2(x) \le 0) \cap (g_3(x) \le 0) \cap (g_4(x) \le 0)\}$$

represents the event that both subsystems of parallel components fail. Figure 2.16 shows the system failure domain of the first-order approximation.

For most system problems, the probability of failure can be estimated even due to the difficulties in estimating ϕ_n, (β, R). For example, let us assume that the

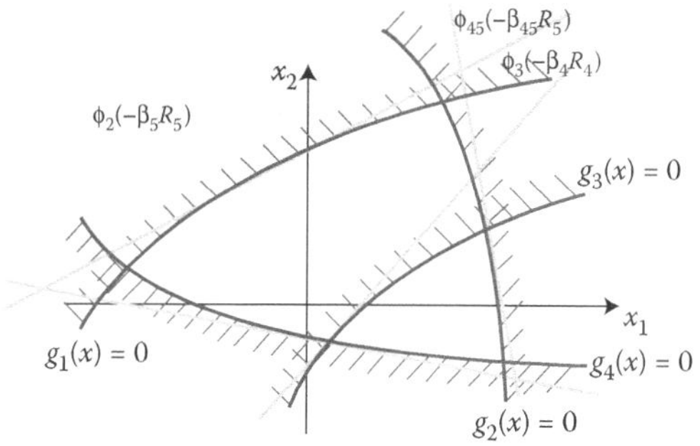

FIGURE 2.16 System failure domain of first-order approximation.

probability of failure of a component reliability analysis is computed as given below:

$$p_f = P(g(x) \leq 0) \tag{2.88}$$

If an inequality event is observed for the component, then the above equation will be modified as

$$p_f = P(g(x) \leq 0 \mid \cap h(x)) \leq 0 \tag{2.89}$$

The above can also be rewritten as

$$p_f = \frac{P(g(x) \leq 0 \cap h(x) \leq 0)}{p(h(x) \leq 0)} \tag{2.90}$$

Thus, the additional information provided by the inequality event can be incorporated into the reliability assessment as explained in the plausible reasoning as seen in Chapter 1. Information from an equality event can be incorporated into a reliability assessment as follows:

$$\begin{aligned} p_f &= P(g(x) \leq 0 \mid h(x) = 0) \\ &= \frac{P(g(x) \leq 0 \cap h(x) = 0)}{P(h(x) = 0)} \\ &= \frac{(\partial / \partial \varepsilon) P(g(x) \leq 0 \cap h(x) = 0)}{(\partial / \partial \varepsilon) P(h(x) - \varepsilon \leq 0)} \\ \lim_{c \to 0} \frac{\partial}{\partial \varepsilon} P(h(x) - \varepsilon \leq 0) &= \frac{P(h(x) - \varepsilon - d\varepsilon \leq 0) - P(h(x) - \varepsilon \leq 0)}{d\varepsilon} \\ &= \frac{P(\varepsilon \leq h(x) \leq \varepsilon + d\varepsilon)}{d\varepsilon} \end{aligned} \tag{2.91}$$

In this case, probability sensitivities with respect to the dummy parameter ϵ, for the parallel system are required.

APPENDIX A: TUTORIALS AND SOLUTIONS

1. Is it possible to find a general formula for $p(C \mid A + B)$, analogous to the generalized sum rule, from the product rule and sum rules? If so, derive it; if not, explain why it cannot be done.

   Solution:
   Associative and commutative rules of a logical product require that the relation should take the form:

$$p(AB \mid c) - p(A \mid BC) p(B \mid C) = p(B \mid AC) p(A \mid C) \tag{A2.1}$$

This is a necessary condition for consistency. This remains consistent for joint propositions also. In the above case, if A is certain, for a given C, then in logical environment produced by knowledge of C, propositions AB and B are the same and have the same plausibility of $AB|C = B|C$ and also $A|BC = A|C$ because A is already certain (for given C) and any other additional information B, which does not contradict C, is still certain and plausible. Hence,

$$\text{Equation A2.1} \Rightarrow p(B \mid C) = p(A \mid C)p(B \mid C) \tag{A2.2}$$

Now the plausibility of $p(C|A + B)$, plausibility of C is ascertained based on the logical sum of propositions (A,B). This disjunctive operation is true at least if one of them is true. We have already established that two essential rules, namely the product and sum rule can be expressed through conjunction AB and negation A, which are adequate for constructing all logical functions.

Hence it is not possible to find a general formula for $p(C|A + B)$, analogous to the generalized sum rule, from the product and sum rules.

2. Suppose we have a set of propositions A_1, A_2, …, A_n which on information X are mutually exclusive: $p(A_iA_j|X) = p(A_i|X)\delta_{ij}$. Show that $p(C(A_1 + A_2 + \cdots + A_n)X)$ is a weighted average of the separate plausibility $p(C|A_1X)$:

$$p(C(A_1 + A_2 + \cdots + A_n)X) = p(C(A_1X + A_2X + \cdots + A_nX))$$
$$= \frac{\sum_i p(A_iX \mid X)p(C \mid A_iX)}{p(A_iX)} \tag{A2.3}$$

Solution:
The two basic rules we have are as follows:

$$p(AB \mid C) = p(A \mid C)p(B \mid AC) = p(B \mid C)p(A \mid BC) \tag{A2.4}$$

$$p(A \mid B) + p(\bar{A} \mid B) = 1 \tag{A2.5}$$

Say

$$p(AB \mid C) = p(A \mid C)p(B \mid AC) \text{ then } p(B \mid AC)$$
$$= \frac{p(AB \mid C)}{p(A \mid C)} \text{ and } (A \mid \overline{BC}) = \frac{p(A\bar{B} \mid C)}{p(\bar{B} \mid C)} \tag{A2.6}$$

$$\text{if } C \equiv A \Rightarrow B, \frac{A \text{ is true}}{B \text{ is true}}; \frac{B \text{ is false}}{A \text{ is false}} \tag{A2.7}$$

This states that $p(AB \mid C) = p(A \mid C)$ and $p(A\overline{B \mid C}) = 0$. Now by substituting these values in Equation A2.6, we get

$$p(B \mid AC) = 1 \text{ and } p(A \mid \bar{B}C) = 0$$

Now

$$A \Rightarrow B, \frac{B \text{ is true}}{A \text{ becomes more plausible}}$$

then

$$p(A \mid C)p(B \mid AC) = p(B \mid C)p(A \mid BC)$$

$$p(A \mid BC) = p(A \mid C)\frac{p(B \mid AC)}{p(B \mid C)} \tag{A2.8}$$

For

$$A \Rightarrow B, \frac{A \text{ true}}{B \text{ true}};$$

$$\frac{A \text{ false}}{B \text{ false}} \text{ means } p(B \mid AC) = 1 \text{ and } p(B \mid C) \leq 1; p(A \mid BC) \geq p(A \mid C) \tag{A2.9}$$

also for

$$\frac{A \text{ is false}}{B \text{ becomes less plausible}} \text{ means that} \tag{A2.10}$$

Also

$$p(B \mid \bar{A}C) = p(B \mid C)\frac{p(\bar{A} \mid BC)}{p(\bar{A} \mid C)} \tag{A2.11}$$

Since

$$p(A \mid BC) \geq p(A \mid C); p(\bar{A} \mid BC) \leq p(\bar{A} \mid C) \tag{A2.12}$$

Hence

$$p(B \mid \bar{A} \mid C) \leq p(B \mid C) \tag{A2.13}$$

Thus if A is true, B becomes more plausible and takes the form

$$p(B \mid AC) > p(B \mid C) \text{ and } p(A \mid BC) > p(A \mid C) \tag{A2.14}$$

Also

$$p(A+B \mid C) = p(A \mid C) + p(B \mid C) - p(AB \mid C)$$

Extending this,

$$p(A_1 + A_2 + A_3 \mid B) = p(A_1 \mid B) + p(A_1 + A_2 \mid B) - p(A_1A_3 + A_2A_3 \mid B)$$

By considering $(A_1 + A_2)$ as one proposition, we get

$$\begin{aligned} p(A_1 + A_2 + A_3 \mid B) &= p(A_1 \mid B) + p(A_2 \mid B) + p(A_3 \mid B) \\ &- p(A_1A_2 \mid B) - p(A_2A_3 \mid B) - p(A_3A_1 \mid B) + p(A_1A_2A_3 \mid B) \end{aligned} \tag{A2.15}$$

Since the propositions $(A_1, A_2,\ldots,A_n)$ are exclusive no two terms of them can be true. Hence, all the terms except first three terms are cancelled. Thus,

$$p(A_1 + A_2 + A_3 \mid B) = \sum ip(A_i \mid B) \tag{A2.16}$$

$$p(A_1 + A_2 + A_n \mid X) = \sum ip(A_i \mid X)$$

Therefore,

$$p(C \mid (A_1 + A_2 + \cdots + A_n)X) = p(C \mid A_1X + A_2X + A_3X + \ldots + A_nX) = p\left(C \mid \sum A_iX\right)$$

$$= \frac{\sum ip(A_i \mid X)p(C \mid A_iX)}{p(A_i \mid X)} \tag{A2.17}$$

3. As soon as we have numerical values for $a = p(A \mid C)$ and $b = p(B \mid C)$ the product and sum rule place some limits on the possible values for their disjunction and conjunction. Supposing that $a \le b$ show that the probability for the conjunction cannot exceed that of the least probable proposition: $0 \le p(AB \mid C) \le a$ and the probability for the disjunction cannot be less than that of the most probable propositions $b \le p(A + BC) \le 1$

Solution:

$$\text{If } C \equiv A \Rightarrow B, \text{ then } \frac{A \text{ is true}}{B \text{ is true}}; \frac{B \text{ is false}}{A \text{ is false}} \text{ states that}$$

$$p(A \mid BC) > p(A \mid C) \text{ and } p(B \mid AC) > p(B \mid C) \tag{A2.18}$$

Provided that the additional information does not contradict the statement, we get

$$p(AB \mid C) = p(A \mid C)p(B \mid AC)$$

For

$$p(A \mid C) = a \text{ and } p(B \mid AC) > p(B \mid C) > B \tag{A2.19}$$

and $a > b$. Therefore, the least possible proposition is $0 \leq p(AB \mid C) \leq a$.
Similarly,

$$p(A + B \mid C) = p(A \mid C) + p(B \mid C) - p(AB \mid C) \tag{A2.20}$$

For $0 \leq p(AB \mid C) \leq a$ and $a < b$, $p(A + B \mid C)$ will be greater than $p(B \mid C)$, that is, b. Hence, the disjunction cannot be less than the probable proposition $b \leq p(A + B \mid C) \leq 1$.

4. The M urns are numbered 1 to M and M balls are also numbered 1 to M. The balls are now thrown into them, one in each turn. If the ball and urn numbers are the same, there is a match. Show that the probability for at least one match is $h = \sum_{k=1}^{M}(-1)^{k+1}/k!$

   As $M \to \infty$, this converges into $1 - 1/e = 0.632$. Note that however large M is, there remains a significant probability for no match at all.

   Solution:

   $$p\text{ (atleast one match)} = \sum_{i=1} p(A_{ij} \mid X) = 1/M \quad \{\text{to have a match, } i = j\}$$

   $$p\text{ (no match at all)} = p(\bar{A}_{ij} \mid X) = 1 - 1/M$$

   For $M = 3$, probability of at least having one match shall be 1/3.

5. N balls are tossed into M urns, there are evidently M^N ways of doing it. If they are considered to be equally likely, what is the probability that each urn can receive at least 1 ball?

   Solution:
   The probability that each urn receives at least 1 ball could be $1/M^N$. For example, if there are 3 urns and 3 balls, then probability of at least each urn receiving 1 ball is 1/27.

6. A total of 100 concrete cubes are tested for the compressive strength and the test data are as given below: Draw the probability density function (pdf) and cdf for the given data (Figure A2.1).

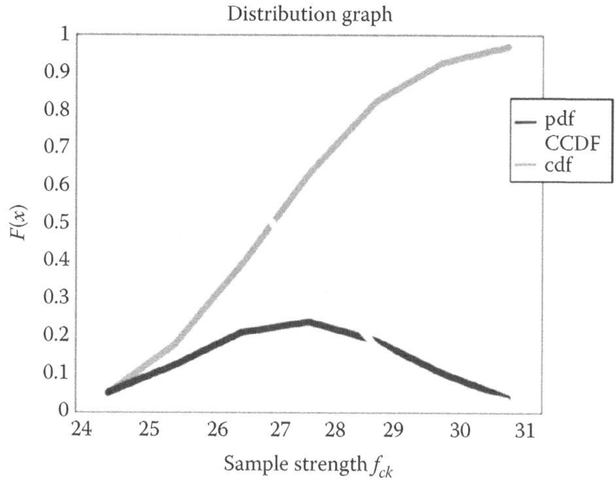

FIGURE A2.1 cdf and pdf for the example problem. CCDF—Complementary cumulative distribution function.

f_{ck} (N/mm²)	Frequency of the Class
24–25	7
25–26	15
26–27	20
27–28	25
28–29	15
29–30	10
30–31	8

Solution:

Mean $f_{ck} = (7 \times 24.5 + 15 \times 25.5 + 20 \times 26.5 + 25 \times 27.5 + 15 \times 28.5 + 10 \times 29.5 + 8 \times 30.5)/100 = 27.38$

$$\bar{f}_{ck} = 27.38$$

$$\sigma = \sqrt{((f_{cki} - f_{\overline{ck}})^2/(N-1))} = 1.647$$

$$\text{Covariance} = \frac{\sigma}{f_{ck}} = \frac{1.647}{27.38} = 0.06 < 1$$

$$f_x(x) = \frac{1}{\sigma_x \sqrt{2\pi}} e^{-1/2\left[\frac{(x-\mu_x)^2}{\sigma_x^2}\right]}$$

Though the cubes f_{ck} values vary between 24 and 31 the deviation of each cube from mean is less.

7. Out of 850 manufactured parts of a machinery component, 50 do not confirm to the requisite standards; two parts are taken at random without replacement from the batch to assess the quality check. Find the cumulative mass function.

   Let random variable X which is equal to the non-confirming part in the sample.

   Probability mass function of X

   $P(X = 0) = (800/850) \times (799/849) = 0.884$
   $P(X = 1) = (800/850) \times (50/849) \times 2 = 0.110$
   $P(X = 2) = (50/850) \times (49/849) = 0.003$

   Cumulative mass function

   $F(0) = P(X< =0) = 0.884$
   $F(1) = P(X< =1) = 0.884 + 0.11 = 0.994$
   $F(2) = P(X< =2) = 0.994 + 0.003 = 1$

8. A, B, and C are three people waiting for placements in a company. Only one job vacancy is available. Data say that A has 0.5 chances of that of B; B has 0.667 of that of C. What is the probability that A or B or C shall get the job?

   $P(A) + P(B) + P(C) = 1$; they are mutually exclusive.

   Given that

   $P[A] = (1/2) \times P[B]$
   $P[B] = (2/3) \times P[C]$
   $P[C] = (3/2) \times P[B]$

   Therefore, $(1/2)P[B] + P[B] + (3/2)P[B] = 1$

   $2P[B] = P[B] = 1$
   $P[B] = (1/3) = 33\%$
   $P[C] = 51\%$
   $P[A] = 17\%$

9. Out of three accelerometers available in the lab, find the probability of at least one accelerometer working after 1 year.

   Solution:
   Assigning the probabilities as G for Good, and B for bad, simply by possible combinations, one can assess the required probability. To start with, all the three accelerometers were new and will be working well. Therefore, G is assigned for all. Pick up the probability of at least one of them is in working condition.

Accelerometer 1	Accelerometer 2	Accelerometer 3
G	G	G
G	G	B
G	B	G
B	G	G
G	B	B
B	G	B
B	B	G
B	B	B

P (1 accelerometer after 1 year) = 3/8

10. In the test conducted on concrete mix design, it is concluded that 30% of coarse aggregate, 25% of fine aggregate, 25% of cement, and 20% of water generally control the overall strength. These data were arrived at based on experiments conducted in the lab. However, it is understood that not all samples tested were reliable and therefore one cannot rely on the statement made for any further assessment. It is now necessary to examine the adequacy of the test results; a sample having the following data is to be assessed. Course A = 27%; FA = 10%; Cement = 30%; and water = 50%. What is the reliability of information from the sampling on one of the random samples?

    Solution:
    We shall use Total Probability Theory, to solve this problem:

    $p[A \mid B] = p[A]p[B \mid A]$

    $P[\text{conc} \mid B] = p[\text{conc}] \times p[B \mid A] + p[\text{conc}] \times p[B \mid CA] + p[\text{conc}] \times p[\text{conc} \mid \text{cement}] + p[\text{conc}] \times p[\text{conc} \mid \text{water}]$

    $= 0.3 \times 0.27 + 0.25 \times 0.1 + 0.25 \times 0.3 + 0.2 \times 0.5$

    $= 0.281$

11. A sample of study was conducted to estimate the *in situ* strength of offshore structures. The sampling was done only on concrete platforms. Destructive testing was followed and core cuttings were made. Test standards suggest that the concrete core should be rejected if at least one out of four is not satisfactory. Estimate the probability that the core cut from the platform is accepted if 100 samples are available and gives poor results

    Solution:
    Let $E(i)$ be the probability of finding that the sample is accepted. For the four core being cut,

$$P[\text{acceptance}] = \frac{95}{100} \times \frac{94}{99} \times \frac{93}{98} \times \frac{92}{97} = 0.812$$

For 50 samples,

$$P[\text{acceptance}] = \frac{45}{50} \times \frac{44}{49} \times \frac{43}{48} \times \frac{42}{47} = 0.65$$

For 25 samples,

$$P[\text{acceptance}] = \frac{20}{25} \times \frac{19}{24} \times \frac{18}{23} \times \frac{17}{22} = 0.38$$

For 200 samples, $P[200] = 0.9$.

APPENDIX B: APPLICATION PROBLEMS

In this section, we shall discuss a few examples where reliability (or safety study) is applied. In offshore compliant structures like TLPs, where buoyancy exceeds weight by design, station-keeping of the platform under extreme loads is an interesting problem. The following example discusses one such numerical case. Though the reliability index is not computed as an outcome of the problem, the safety of the platform under extreme loads is examined, which is a part of the reliability study.

B.2.1 Mathieu's Stability

This example presents the dynamic analysis of tethers and TLPs considering the linearly varying tension along the tether length. The modal analysis considers a linear cable equation for tether modeling subjected to tension, which varies along its length. A Mathieu stability analysis is then performed for TLPs of different shapes and different water depth of 527.8, 872, and 1200 m, respectively, to obtain the amplitudes of tether vibrations. Unstable modes of vibration are also verified. The resultant modal forms for the tether's dynamic model are then obtained in the form of Bessel's function. In the present example, the following assumptions are considered: (i) static tension varies linearly along the length due to the effect of its own submerged weight; (ii) end conditions are considered as simply supported; (iii) current effect is not included; and (iv) flexural rigidity of the tether is neglected since it is very slender. Once the tether's flexural rigidity is neglected, its lateral movement is governed by the following equation:

$$M\frac{\partial^2 y}{\partial t^2} - \frac{\partial}{\partial x}\left(T(x)\frac{\partial y}{\partial x}\right) + Bv\left|\frac{\partial y}{\partial x}\right|\frac{\partial y}{\partial t} \tag{B2.1}$$

where M is the tether's mass per unit length (physical plus added masses), B_v the quadratic fluid damping coefficient (viscous), and $T(x)$ is the total tension. The total tension, acting along the tether is composed by the static tension ($T_0(x)$), resulting from the platform pretension (P) and varying along the tether's length due to the action of its submerged weight is given by

$$T_0(x) = P + \mu g(L - x) \tag{B2.2}$$

where μ is the physical mass per unit length, L the tether's length. The dynamic tension ($T_d(x)$), imposed by the hull's first-order heave motion induced by a monochromatic wave with frequency (ω) is given by

$$T_d(x) = -S\cos(\omega t) \tag{B2.3}$$

where S is the wave-induced tension amplitude, leading to a total tension, which is given by

$$T(x) = P + \mu g(L - X) - S\cos(\omega t) \tag{B2.4}$$

The dynamic equation for the tether's free lateral vibration is given by

$$M\frac{\partial^2 y}{\partial t^2} - \frac{\partial T(x)}{\partial x} = 0 \tag{B2.5}$$

Assuming the lateral motion of the nth natural mode as follows:

$$y_n(x,t) = f_n(t)X_n(x) \tag{B2.6}$$

and substituting Equation B2.6 into Equation B2.1, we get

$$\left\{\left[\frac{P}{M} + \frac{\mu g(L - x)}{M}\right]X_n\right\} + \bar{\omega}_n^2 X_n = 0 \tag{B2.7}$$

where $\bar{\omega}_n$ is the natural frequency of the nth mode. As shown in Equation B2.7, the dynamic model leads to a classical Sturm–Liouville problem. By introducing a new variable η, which is given by

$$\eta = \sqrt{1 + \frac{\mu g(L - x)}{P}} \tag{B2.8}$$

Equation B2.7 can be rewritten in the following form:

$$\eta^2 X_n + \eta X_n + 4\beta_n \eta^2 X_n = 0$$

$$\beta_n^2 = \frac{PM}{(\mu g)^2}\bar{\omega}_n^2 \tag{B2.9}$$

The above equation is identified as a modified Bessel equation, where the solution is given in the form of the Bessel function ($J_0 y_0$):

$$Xn(n) = C_1 J_0(2\beta_n \eta) + C_2 Y_0(2\beta_n \eta)$$

The constants (C_1,C_2) are determined by considering the appropriate boundary condition, such as

$$\begin{cases} X_n(\eta|_{r=0}) = 0 \\ X_n(\eta|_{r=1}) = 0 \end{cases}$$

The resulting model forms for the tether's dynamic model are then obtained as

$$\begin{aligned} X_n(x) = J_0\left(2\beta_n\left[1+\frac{\mu g(L-x)}{P}\right]^{1/2}\right) \\ -\frac{J_0(2\beta_n)}{Y_0(\beta_n)} Y_0\left(2\beta_n\left[\frac{\mu g(L-x)}{P}\right]^{1/2}\right) \end{aligned} \tag{B2.12}$$

Frequencies are given by

$$\bar{\omega}_n = \mu g\sqrt{\frac{1}{PM}}\beta_n \tag{B2.13}$$

The values of β_n are determined numerically as the solution of the characteristic equation are given by

$$J_0\left(2\beta_n\sqrt{1+\frac{\mu gL}{P}}\right) y_0(2\beta_n) - y_0\left(2\beta_n\sqrt{1+\frac{\mu gL}{P}}\right) j_0(2\beta_n) = 0 \tag{B2.14}$$

Substituting Equation B2.12 into Equation B2.1 and applying Galerkin's variation method (multiplying it by $X_n(x)$ and integrating it along the tether's length) and considering the following coefficients:

$$\begin{aligned} I_1 = \int_0^L X_n^2(x)dx I_2 = \int_0^L \frac{dX_n}{dx} X_n\, dx I_3 = \int_0^L \frac{d^2X_n}{dx^2} X_n\, dx \\ I_4 = \int_0^L \frac{d^2X_n}{dx^2} X_n x\, dx I_5 = \int_0^L X^3(x)dx \end{aligned} \tag{B2.15}$$

Equation B2.1 can now be expressed in the following form:

$$\frac{d^2 f}{d\tau^2} + (\delta - q\cos(2\tau))f + c\left|\frac{df}{d\tau}\right|\frac{df}{d\tau} = 0 \tag{B2.16}$$

where τ is the dimensionless time variable ($2\tau = \omega t$).

The tether's dynamics are hence represented by a Mathieu parametric equation with parameters given by

$$\delta_n = \frac{4}{M\omega^2}\left\{\mu g \frac{(I_2 + I_4)}{I_1}\right\} - (P + \mu g L)\frac{I_3}{I_1} q_n$$
$$= -\frac{2S}{M\omega^2}\frac{I_3}{I_1} C_n = \frac{B_v}{M}\frac{I_5}{I_1} \tag{B2.17}$$

The dynamic analysis is now performed assuming an induced dynamic tension amplitude(s) equivalent to 60% of the nominal static tension (as an extreme case that can induce failure). Stability analysis is carried out for TLPs namely TLP_1 and TLP_2 (three-legged) and Auger TLP (four-legged), respectively. The plan of three-legged TLP is shown in Figure B2.1.

The stability analysis to be performed can be illustrated graphically by the Mathieu stability chart (shaded regions are unstable) (Figure B2.2).

Geometric properties of TLP taken for the study are given in Table 2.2. From Table 2.3 (for TLP_1) and Table 2.4 (for TLP_2) it is seen that the parameters δ and q fall below the shaded region of the Mathieu stability chart, indicating that TLP_1 and TLP_2 are stable in first mode for C_m 1.5 and 2.0, respectively.

Figures B2.3 through B2.6 show the generalized displacement amplitude for TLP_2 with C_m 1.5 and 2.0, respectively. It is seen from these figures that the amplitude is decaying exponentially and becomes nearly zero at 60s. It is also seen that for

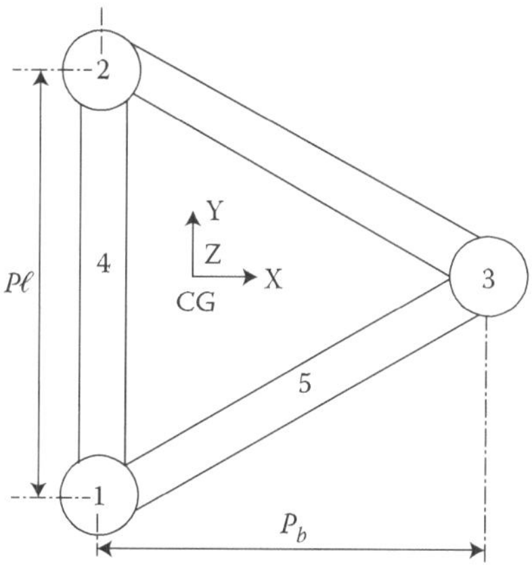

FIGURE B2.1 Plan of TLP, considered for the example problem.

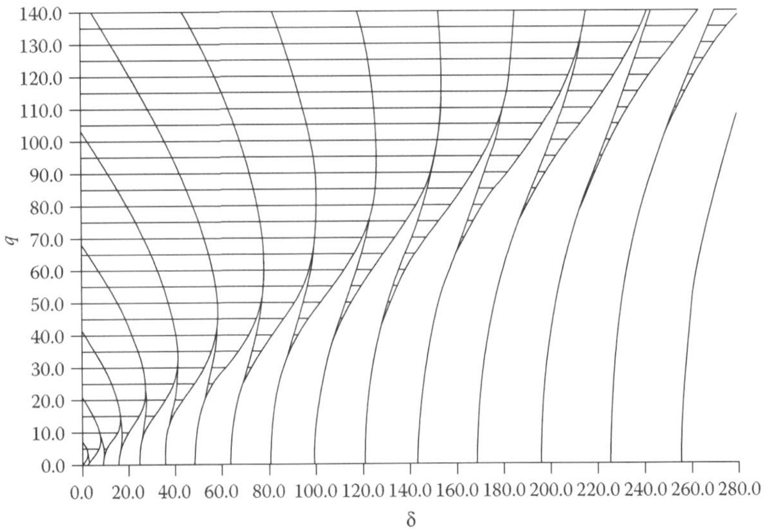

FIGURE B2.2 Mathieu's stability chart.

TABLE 2.2
Geometric Properties of Tension Leg for Stability Analysis

Description	TLP_1	TLP_2	Auger TLP
P (pretension in one tether) (kN)	12,495.0	21,291.675	9030.0
Tether length (m)	485.0	1166.0	834.0
Water depth (m)	527.8	1200.0	872.0
No. of tethers	12 (three groups)	12 (three groups)	12 (four groups)
External diameter of tether (m)	0.66	0.66	0.66
Thickness of tether wall (m)	0.033	0.033	0.033

TABLE 2.3
Stability Parameters for TLP_1

Description	C_m	q	δ	Condition
TLP_1	1.5	122.906	418.3	Stable
	2.0	135.3428	410.8	Stable

TABLE 2.4
Stability Parameters for TLP_2

Description	C_m	q	δ	Condition
TLP_2	1.5	140.2	361.19	Stable
	2.0	156.36	371.15	Stable

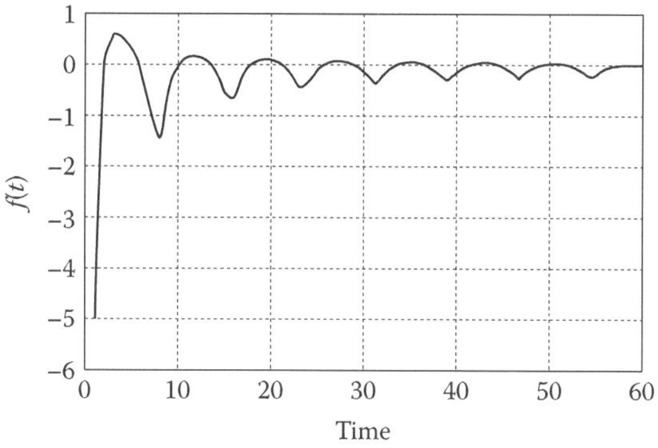

FIGURE B2.3 Generalized displacement amplitude for TLP_1 tethers—first mode ($C_m = 1.5$).

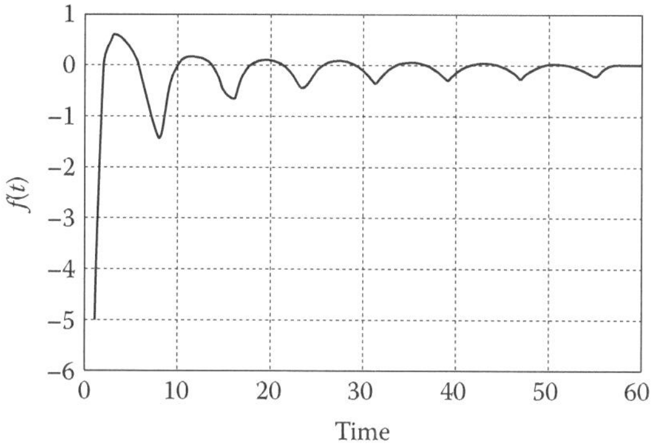

FIGURE B2.4 Generalized displacement amplitude for TLP_1 tethers—first mode ($C_m = 2.0$).

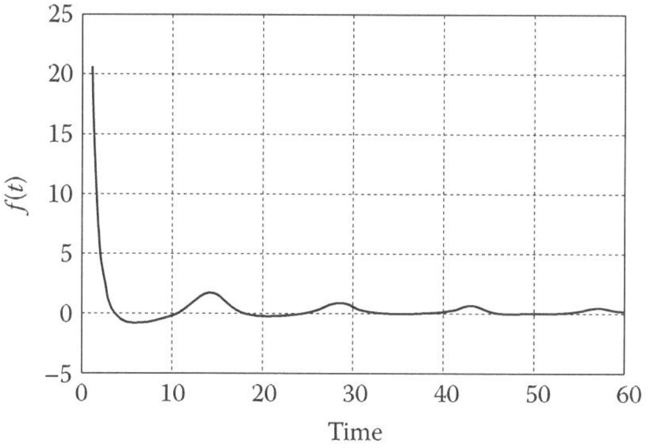

FIGURE B2.5 Generalized displacement amplitude for TLP_2 tethers—first mode ($C_m = 1.5$).

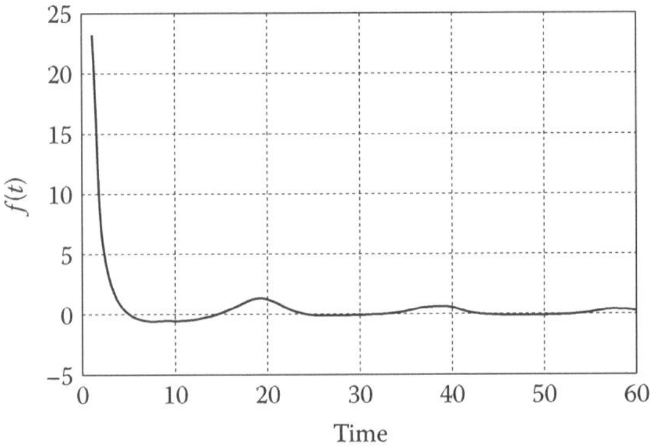

FIGURE B2.6 Generalized displacement amplitude for TLP_2 tethers—first mode (C_m = 2.0).

the increase of C_m from 1.5 to 2.0 (i.e., 33.33%) the stability parameter q increased by 10.9% and parameter δ decreased by 1.8% for TLP_1. Similarly for TLP_2, q increased by 10.34% and δ also increased by 2.7%. With reference to the Mathieu stability chart, this shows that the increase in q with increase in δ move toward the stability zone. Hence, with greater increase in q and marginal decrease in δ, the region will lie in the stability zone. Therefore the increase in C_m, which contributes to the added hydrodynamic mass also increases the stability of the TLP in deep water.

The Auger's operation condition leads to an unstable response under the tether's first mode of vibration as shown in Figures B2.7 and B2.8 for C_m 1.5 and 2.0, respectively.

From Table 2.5 it is seen that the parameters δ and q fall in the shaded region of the Mathieu stability chart indicating that the Auger TLP is unstable in first mode for C_m 1.5 and 2.0, respectively.

It is seen from the figures that the Auger TLP is unstable under the principle mode of vibration as the plot of displacement amplitude does not shown any decay even after a large passage of time. Hence, the influence of tension variation along the

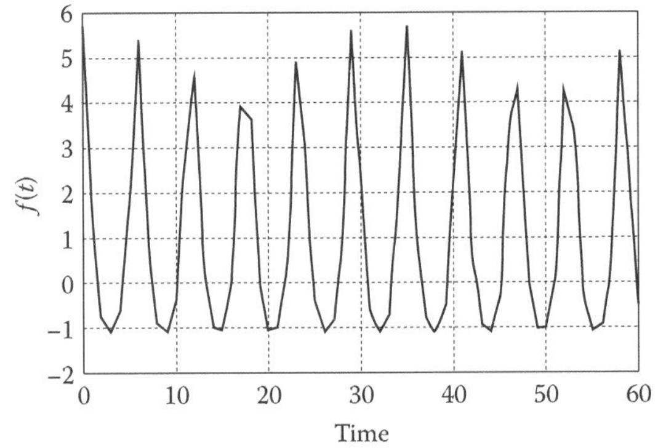

FIGURE B2.7 Generalized displacement amplitude for Auger TLP tethers—first mode (C_m = 1.5).

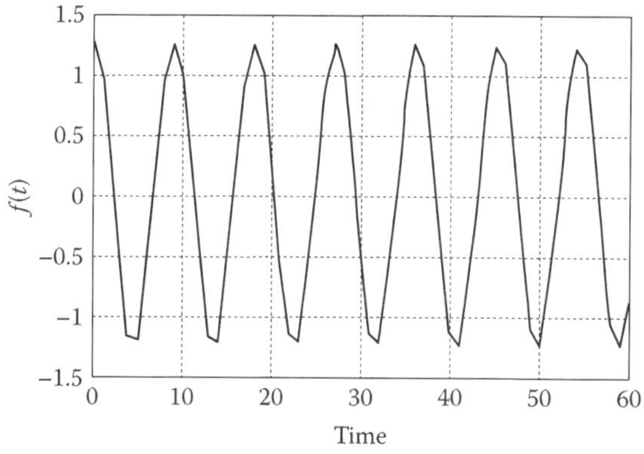

FIGURE B2.8 Generalized displacement amplitude for Auger TLP tethers—first mode ($C_m = 2.0$).

TABLE 2.5
Stability Parameter for Auger TLP

Description	C_m	q	δ	Condition
Auger TLP	1.5	221.8	2.413	Unstable
	2.0	285	1.025	Unstable

tether's length plays an important role in its stability dynamics. It is also seen that triangular TLP for the selected water depth and increased pretension shows good agreement of stability in the selected water depth. Increased pretension shows good agreement of stability in the principle modes of vibration when compared to that of the four-legged Auger TLP.

The example problem highlights the fact that tether tension variation plays an important role in the stability analysis of TLPs. Its consideration is indeed necessary since the water depth increases with deep-water compliant structures. Lesser pretension value, as shown in the Auger TLP lead to an unstable situation under the considered extreme loads. It is also seen that a new geometric form, which is a triangular configuration, enhances stability in the first fundamental mode of vibration.

B.2.2 STABILITY OF TETHERS UNDER DISTINCTLY HIGH SEA WAVES AND SEISMIC EXCITATION

As reliability is to assess satisfactory performance under the given load combination for a certain period of time, the presented example highlights the stability of tethers under distinctly high sea waves and seismic excitations, which are real situations (simulated in the numerical study). It is therefore imperative to understand the dynamic response under such uncertain conditions.

This example study describes a mathematical model of dynamic forces on TLPs caused by distinctly high sea waves in the presence of both horizontal and vertical seismic excitations and discusses the method of analysis. Seismic forces imposed at bottom of each tether as axial forces make tether tension unbalanced when the hull is under offset condition. The tether tension varies nonlinearly under vertical seismic excitation generated using the Kanai–Tajimi ground acceleration spectrum. The following assumptions are made in the example: (i) the platform is considered as a rigid body having six degrees of freedom; (ii) water waves, generated due to the ground motion, are neglected; (iii) hydrodynamic force coefficients namely C_d and C_m are same for pontoons and columns and are independent of wave frequencies; (iv) C_m is considered to vary with water depth. Values of C_m at seabed, mid-depth, and mean sea level (MSL) have been taken as 1.8, 1.58, and 1.5, respectively, and other values have been interpolated by a second-degree polynomial fit; (v) stiffness of the tethers is modeled as

$$\text{stiffness} = \begin{cases} \dfrac{AE}{l}; \text{elongation} \geq 0 \\ 0; \quad \text{elongation} < 0 \end{cases}$$

This implies that stiffness becomes zero when the decrease in tether tension due to motion of the platform is more than its initial pretension and tethers become slack.

The shape of distinctly high sea waves is crucial for a given sea state. Experimental investigations conducted by researchers (see, e.g., Kim et al., 1994) show that the shape of these waves is steep and asymmetric with respect to both horizontal and vertical axes. Kim and Zou (1995) developed a method to generate freak waves in wave tanks based on the design wave spectrum determined numerically resulting also in a highly asymmetric wave profile. In the present example, sea waves are generated as a random time history using the Pierson–Moskowitz (PM) spectrum. Generation of distinctly high sea waves calls for implementation of a nonlinear wave kinematic theory and nonlinear fluid model. Keeping in view that: (i) nonlinear wave theories tend to become overly complicated even when a low-order approximation is used; and (ii) several researchers (see, e.g., Pilotto and Ronalds, 2003; Pilotto et al., 2002, 2003) have successfully simulated such high sea waves using Airy's wave theory and the Morison equation in drag-dominated offshore structures, the same have been used in this study. Water particle kinematics are obtained using Airy's wave theory from a randomly generated sea surface elevation using the PM spectrum. The original PM spectrum, a function of wind velocity is modified as function of modal frequency and later modified again as a function of significant wave height and modal frequency (see, e.g., Michel, 1999). The modified one parameter formula employed in the study is given by

$$S_{\eta\eta}(\omega) = \frac{8.1\times10^{-3} g^2}{\omega^5} \exp\left[-1.25\left(\frac{\omega_m}{\omega}\right)^4\right] \qquad \text{(B2.18)}$$

where g is the acceleration due to gravity, ω_m the modal frequency (taken as 0.46 rad/s [0.07 Hz] in the current study), and $S_{\eta\eta}$ is the power spectral density of wave height. The wave elevation, $\eta(t)$, realized as a discrete sum of many sinusoidal functions with different angular frequencies and random phase angles is given by

$$\eta(t) = \sum_{i=1}^{n} \sqrt{2S_{\eta\eta}(\omega_i)\Delta\omega_i} \cos(k_i x - \omega_i t - \varphi_i) \tag{B2.19}$$

where k_i is the wave number, ω_i are discrete sampling frequencies, $\Delta\omega_i = \omega_i - \omega_{i-1}$, n is the number of data points, and ϕ_i are random phase angles. The generated wave profile is designed to have a peak at a particular time (t_0), which will be distinctly high in comparison to other wave heights and therefore classified as distinctly high sea waves. Figure B2.9 shows the PM spectrum and the sample wave thus generated using Equation B2.19 as seen at $x = 0$ (i.e., the place where waves first meet the TLP). A magnified time history from $t = 4$ s to $t = 14$ s shows that the wave thus generated is a continuous function of time near the chosen time interval as well as illustrates the concave front and convex rear of the generated wave to cause the desired impact on TLPs (Kim et al., 1994). Moreover, the spatially varying high sea wave profile is the wave as seen at the first-most point of the TLP, that is, the point where waves confront the TLP at $x = 0$. From there on, different phases of the same wave profile reach different positions along the TLP as the wave passes away.

The example structure, which is a triangular TLP model, is shown in Figure B2.10. Equation of motion describing dynamic equilibrium between inertia, damping, restoring, and exciting forces can be assembled as follows:

$$[M]\{\ddot{X}\} + [C]\{\dot{x}\} + [k]\{x\} = \{F(\{X\},\{\dot{X}\},\{\ddot{X}\},t)\} \tag{B2.20}$$

where $[M]$ is the mass matrix, $[C]$ the damping matrix, $[K]$ the stiffness matrix, and $\{F\}$ is the hydrodynamic force vector. Structural mass is assumed to be lumped at each degree of freedom, is diagonal in nature and is given by

$$[M] = \begin{bmatrix} M_{11}+M_{a11} & 0 & 0 & 0 & 0 & 0 \\ 0 & M_{22} & 0 & 0 & 0 & 0 \\ 0 & 0 & M_{33}+M_{a33} & 0 & 0 & 0 \\ 0 & 0 & 0 & M_{44} & 0 & 0 \\ M_{a51} & 0 & M_{a53} & 0 & M_{55} & 0 \\ 0 & 0 & 0 & 0 & 0 & M_{66} \end{bmatrix} \tag{B2.21}$$

where $M_{11} = M_{22} = M_{33}$ = total mass of the entire platform, M_{44}, M_{55}, M_{66} = total mass moment of inertia about X, Y, Z axes = Mr_x^2, Mr_y^2, Mr_z^2, respectively, r_x, r_y, and r_z are the radii of gyration about X, Y, and Z axes, respectively, M_{a11}, M_{a33} are added mass terms in surge and heave degrees of freedom, M_{a51}, M_{a53} are added mass moment of inertia due to the additional mass in surge and heave degrees of freedom, respectively. Heave added mass on TLP hull columns is taken equivalent to the mass of a hemispherical volume of water, that is, $M_{a23} = \pi\rho D^3/12$ (Chakrabarti and Hanna, 1990).

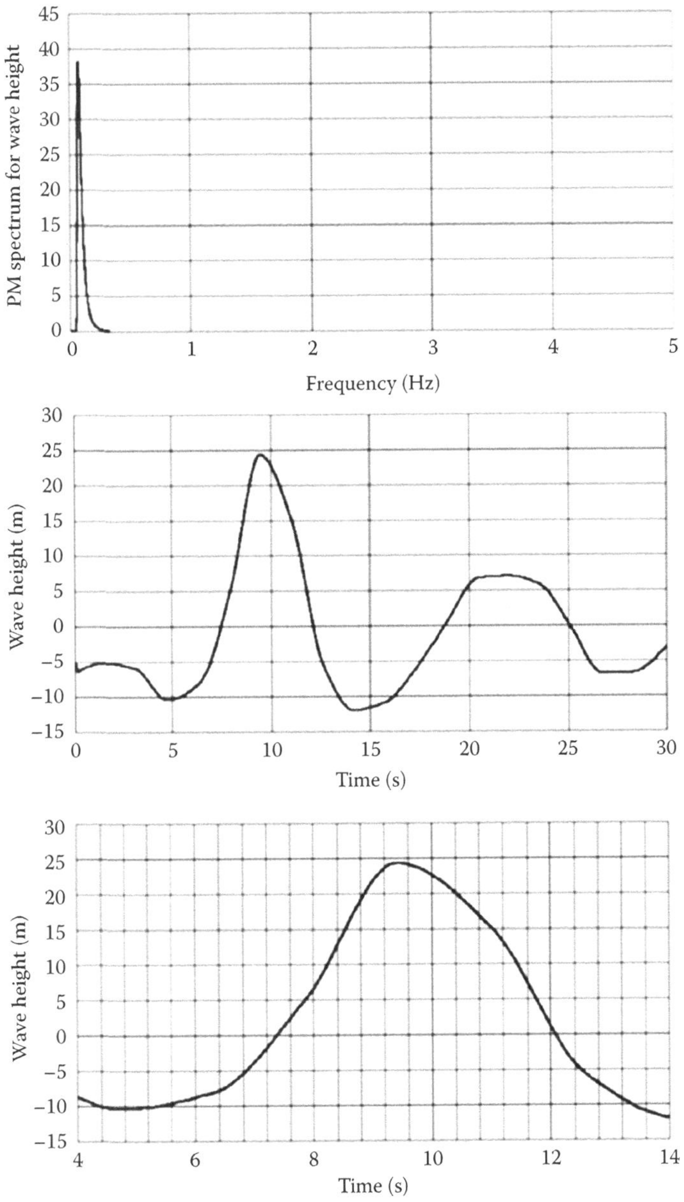

FIGURE B2.9 PM spectrum and sample wave history used in the study.

The presence of off-diagonal terms indicates the contribution of added mass due to hydrodynamic loading.

The contribution of added mass up to MSL has already been considered along with the force vector. The stiffness matrix $[K]$ depends on the tether tension and is response dependent. The stiffness matrix of the triangular TLP is given by

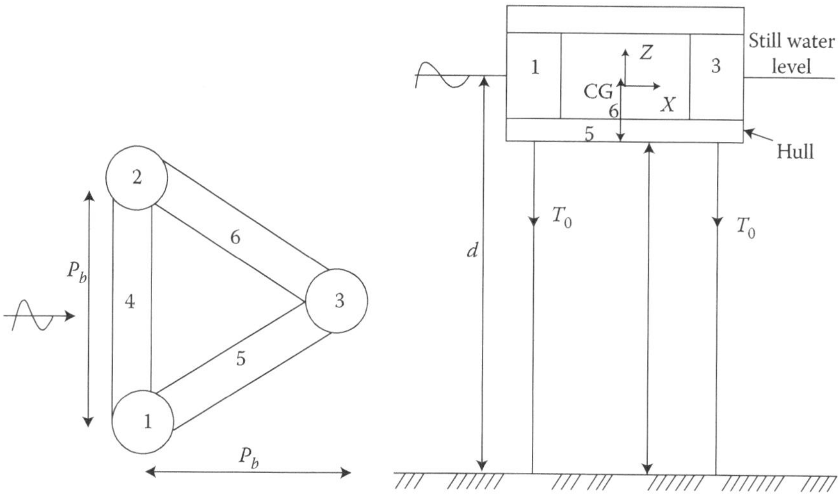

FIGURE B2.10 Example TLP taken for the study.

$$[k]=\begin{bmatrix} K_{11} & 0 & 0 & 0 & 0 & 0 \\ 0 & K_{22} & 0 & 0 & 0 & 0 \\ K_{31} & K_{32} & K_{33} & K_{34} & K_{35} & K_{36} \\ 0 & K_{42} & 0 & K_{44} & 0 & 0 \\ K_{51} & 0 & 0 & 0 & K_{55} & 0 \\ 0 & 0 & 0 & 0 & 0 & K_{66} \end{bmatrix} \quad \text{(B2.22)}$$

Coefficients of the stiffness matrix derived from the first principles can be seen in the literature (see, e.g., Srinivasan Chandrasekaran and Jain, 2002). The damping matrix $[C]$ is assumed to be proportional to the initial values of $[M]$ and $[K]$ and is given by

$$[C]=a_0[M]+a_1[K] \quad \text{(B2.23)}$$

where a_0 and a_1 are, respectively, the stiffness and mass proportional damping constants. The damping matrix given by Equation B2.24 is orthogonal as it permits modes to be uncoupled by eigenvectors associated with the undamped eigenproblem. Damping constants a_0 and a_1 are determined by choosing the fractions of critical damping (ξ_1 and ξ_2) at two different frequencies (ω_1 and ω_2) and solving simultaneous equations for a_0 and a_1.

$$a_0 = 2(\xi_2\omega_2 - \xi_1\omega_1)/(\omega_2^2 - \omega_1^2) \quad \text{(B2.25)}$$

$$a_0 = 2\omega_1\omega_2(\xi_2\omega_2 - \xi_1\omega_1)/(\omega_2^2 - \omega_1^2) \quad \text{(B2.26)}$$

Damping attributable to $a_0[K]$ increases with increasing frequency, whereas damping attributable to $a_1[M]$ increases with decreasing frequency. In this study, the value of these coefficients are obtained using Equations B2.26 and B2.27 by taking the damping ratio $\zeta = 0.05$ in surge and yaw degrees of freedom. Free vibration analysis is performed to find out natural frequencies of the platform corresponding to these degrees of freedom and it is found that damping ratios maintain reasonable values for all the other modes which are contributing significantly to the response (Chopra, 2003). The force vector, at any time instance, is given by

$$\{F(t)\} = \{F_1 F_2 F_3 F_4 F_5 F_6\}^T \tag{B2.27}$$

where F_i will be the force in the ith degree of freedom. The first degree of freedom is surge, second sway, third heave, fourth roll, fifth pitch, and sixth yaw. The dynamic buoyant force is given by

$$F_b(t) = \frac{3}{4}\pi D^2 \rho x(T) \tag{B2.28}$$

Wave forces are modeled using the modified Morison's equation and are given by

$$f(t) = 0.5\rho C_d D\,|\dot{u} - \dot{X}|\,(\dot{u} - \dot{X}) + 0.25\pi D^2 \rho C_m \ddot{u} \pm 0.25\pi D^2 [C_m - 1]\rho \ddot{X} \tag{B2.29}$$

where $(\dot{u} - \dot{X})$ is the instantaneous relative velocity in the considered direction, $\dot{X}$, $\ddot{X}$ are structural velocity and acceleration, respectively, D the diameter of the cylinder, ρ the density of sea water, and C_m, C_d are inertia and drag coefficients, respectively. The water particle kinematics ($\dot{u}$, $\ddot{u}$, $\dot{v}$, $\ddot{v}$) are obtained using Airy's wave theory from a randomly generated sea surface elevation (η) using the PM spectrum. The force vector is updated at every time instance for its response dependence and to account for variable submergence effects. This makes the solution procedure iterative, which is solved using Newmark's β time integration procedure. In addition to hydrodynamic forces arising from distinctly high sea waves generated by Equation B2.18, variations caused in tether tension due to vertical and horizontal seismic excitations are also accounted for. Dynamic tether tension (ΔT) used to update coefficients of stiffness matrix is given by

$$\Delta T = \frac{AE}{l}(X(T) - X_g(t)) \tag{B2.30}$$

where $x(t)$ is the instantaneous response vector of the TLP and $x_g(t) = \{x_{1g}(t), 0, x_{3g}(t), 0, 0, 0, 0\}^T$ is the ground displacement vector. x_{1g} is the horizontal ground displacement (in the surge degree of freedom) and x_{3g} is vertical ground displacement (in the heave degree of freedom). These random ground motions are generated using the Kanai–Tajimi ground spectrum, the one-sided power spectral density of which is given by

$$S_{\ddot{x}_g\ddot{x}_g}(\omega) = \frac{\omega_g^4 + 4\zeta_g^2\omega_g^2\omega^2}{(\omega_g^4 - \omega^2)^2 + 4\zeta_g^2\omega_g^2\omega^2} \tag{B2.31}$$

$$S_0 = \frac{2\zeta_g\sigma_g^2}{\pi\omega_g(1+4\zeta_g^2)} \tag{B2.32}$$

where ω_g and ζ_g are the natural frequency and damping ratio of the oscillator determined by the characteristics of the local earth surface layer, σ_g^2 is the variance of ground acceleration. The intensity (S_o) of excitation is determined by the strength of seismic waves. The three parameters of the Kanai–Tajimi model namely ω_g, ζ_g, and s_0 are estimated from representative earthquake records by means of statistical estimation process. An artificial earthquake is generated to match peak ground velocity of a recent earthquake that occurred in the Gulf of Mexico (M 5.8 magnitude occurred in the Gulf of Mexico, approximately 250 miles WSW of Anna Maria, Florida on 10th September 2006 at 14:56:07 [Coordinated Universal Time] epicentered 26.34N, 86.57W). The signal thus generated is appropriate as, for example, Mars TLP, operating in the Mississippi Canyon Block, is also located in the Gulf of Mexico. The actual earthquake had a peak ground acceleration and velocity of 0.25*g* and 0.29 m/s whereas the artificial one generated is reasonably close having these values as 0.25–0.39*g* and 0.20–0.30 m/s, respectively. The frequency of ground motion, ω_g, is set as 2.5 Hz as reported for firm ground (see, e.g., Nigam, 1983) and other characteristics are manipulated to obtain ground acceleration and velocity closer to the earthquake that occurred in the Gulf of Mexico. Figure B2.11 shows the Kanai–Tajimi spectrum for vertical ground acceleration considered in the analysis and Figure B2.12 shows a sample time history. Change in tether tension given by Equation B2.31 updates stiffness coefficients and thus incorporates the influence of seismic excitation in the analysis. Force vector comprising of two components namely: (1) Forces due to hydrodynamic loading

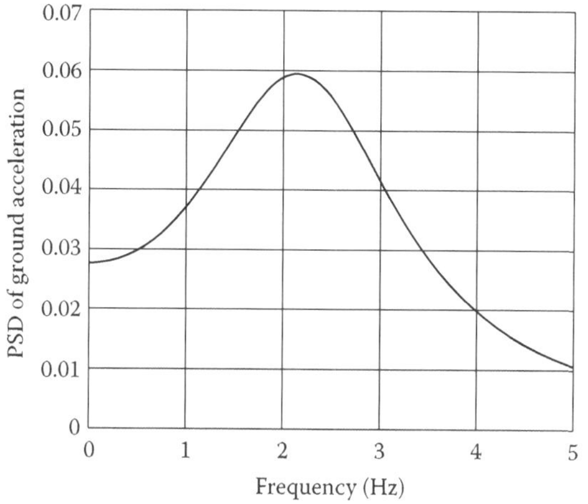

FIGURE B2.11 PSD of the vertical ground acceleration.

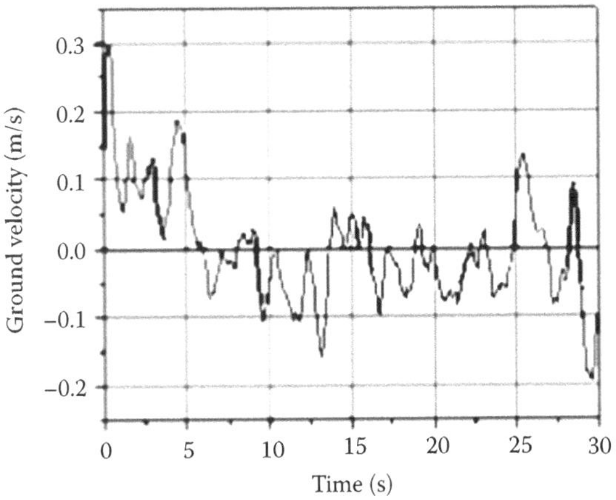

FIGURE B2.12 Time history of ground velocity.

arising from high sea waves; as well as (2) variation in tether tension altering dynamic equilibrium and thus inducing additional inertial forces exclusively caused by vertical and horizontal seismic excitation, is then resolved in all active degrees of freedom.

The geometric properties of typical three square TLPs at different water depths are taken from the literature (see, e.g., Srinivasan Chandrasekaran and Jain, 2002a) and listed in Table B2.1. For arriving at the example TLP, which is of triangular geometry, equations of equilibrium applied for static sea conditions are given by

$$F_B = 4(T_o) + W \tag{B2.33}$$

$$F_B = 3(T_o) + W \tag{B2.34}$$

TABLE B2.1
Geometrical Properties of Example TLPs (Square)

Property	TLP_1	TLP_2	TLP_3
Weight (kN)	209,500.00	330,000.00	370,000.00
F_B (kN)	334,000.00	520,000.00	625,500.00
T_0 (kN)	124,500.00	190,000.00	255,500.00
Tether length, l (m)	471.00	568.00	1,166.00
Water depth, d (m)	500.00	600.0	1,200.00
CG above Keel (m)	26.60	28.50	30.31
AE/l (kN/m)	58,060.00	82,000.00	45,080.00
Plan dimension (m)	92.50	78.50	83.50
D (m)	14.20	17.00	18.80
r_x and r_y (m)	29.15	35.10	35.10
r_z (m)	32.10	42.40	42.40

Note: TLP indicates tension leg platform; CG, center of gravity.

where F_B is the buoyant force, T_o the initial pretension in each tether, and W is the weight of the platform. It can be seen that the example TLP of triangular geometry has the same initial tether tension as that of the square. This results in reduced total tether pretension and therefore weight is increased by keeping the buoyant force same. Coefficient of inertia, C_m is interpolated for the entire water depth with a second-degree polynomial, which is given by

$$c_m(y) = P_1 \cdot y^2 + P_2 \cdot y + P_3 \quad \text{(B2.35)}$$

where y is water depth measured from the seabed. The values of coefficients of Equation B2.36 are given in Table B2.2. Table B2.3 shows the structural time period and frequencies as well for the three triangular TLPs considered in the study. It can be seen that the heave frequency is closer to that of the high sea waves generated (whose modal wave frequency is 0.46 Hz) to notice the bear resonant response situation. The equation of motion is solved in the time domain by employing Newmark's integration scheme by taking $\alpha = 0.25$ and $\delta = 0.5$ (Bathe and Wilson, 1987). The solution procedure incorporates all time-dependent nonlinearities namely: (1) change in stiffness coefficients due to the change in tether tension caused due to seismic excitations; (2) added mass due to variable submergence effect; (3) set-down effect; as well as (4) evaluation of hydrodynamic forces at instantaneous displaced positions of the system.

Dynamic tether tension is one of the important response characteristics of TLP that is influenced by the considered environmental loads as ground motion is

TABLE B2.2
Coefficient for Interpolation of C_m

Triangular TLPs	P_1	P_2	P_3
TLP_1	1.120×10^{-6}	-1.160×10^{-3}	1.8
TLP_2	7.778×10^{-7}	-9.667×10^{-4}	1.8
TLP_3	1.944×10^{-7}	-4.833×10^{-4}	1.8

Note: TLP indicates tension leg platform.

TABLE B2.3
Natural Wave Periods and Natural Frequencies of Triangular TLPs

TLP	Natural Wave Period (s)			Natural Frequency (Hz)		
Nomenclature	Surge	Heave	Pitch	Surge	Heave	Pitch
TLP_1	83.33	1.92	1.960	0.0120	0.5208	0.5102
TLP_2	97.09	1.92	2.060	0.0103	0.5208	0.4854
TLP_3	131.58	3.11	3.120	0.0076	0.3215	0.3205

TABLE B2.4
Response Summary of Triangular TLPs

Description	TLP_1	TLP_2	TLP_3
	Time History Response		
Heave (m)	1.2034	0.7384	1.1139
Pitch (rad)	0.002	0.0025	0.0011
Surge (m)	0.8683	0.8211	0.8655
	PSD Peaks at		
Heave (Hz)	0.0, 0.706, 1.588	0.0, 0.686, 1.588	0.0, 686, 1.588
Pitch (Hz)	0.039, 0.608, 1.588	0.039, 0.569, 1.588	0.02, 0.608, 1.588
Surge (Hz)	0.039, 1.588	0.039, 1.588	0.02, 1.588
Dynamic tether tension variation (kN)	81,455.95	79,051.85	51,898.73
Change in tether tension (%)	65.43	41.606	20.313
Strain in tether (%)	0.298	0.146	0.109

Note: Model wave frequency = 0.07 HZ, ground motion frequency = 2.5 Hz. TLP indicates tension leg platform; PSD, power spectral density.

indirectly considered in the analysis by superimposing the change in tether tension with respect to time. It is interesting to note that this variation, directly superimposed in the stiffness coefficients of TLP, change their response behavior considerably. Table B2.4 shows the response summary of three triangular TLPs considered for the study. Figure B2.13 shows the response time history under seismic forces and distinctly high sea waves.

The percentage change in tether tension is computed as follows: For example, in case of TLP_1, maximum heave response is 1.2034 m and surge response at this time instance is 0.7212 m.

$$\text{Change in tether length} \quad \Delta l = \sqrt{(1.2034)^2 + (0.7212)^2} = 1.4029\,\text{m}$$

$$\text{Additional dynamic tether tension} = \Delta l * \frac{AE}{l} = 1.4029 \times 58{,}060 = 81{,}455.95\,\text{kN}$$

% change in tether tension = additional dynamic tether tension/initial tether tension = 81,455.95/124,500 amounting to 65.43%, as stated in Table B2.4. Similarly, the percentage change in tether tension is 41.61% and 20.31% for TLP_2 and TLP_3, respectively. To show that the tethers have not yielded, the strain in the tethers are also computed and tabulated. It can be seen that dynamic tether tension variation is reduced considerably with increase in water depth but this variation is neither proportional to change in water depth nor initial tether tension. Figures B2.14 and B2.15 are presented in three sets for all three triangular TLPs namely: (i) figure showing response time histories in heave, pitch, and surge degrees of freedom, (ii) figure

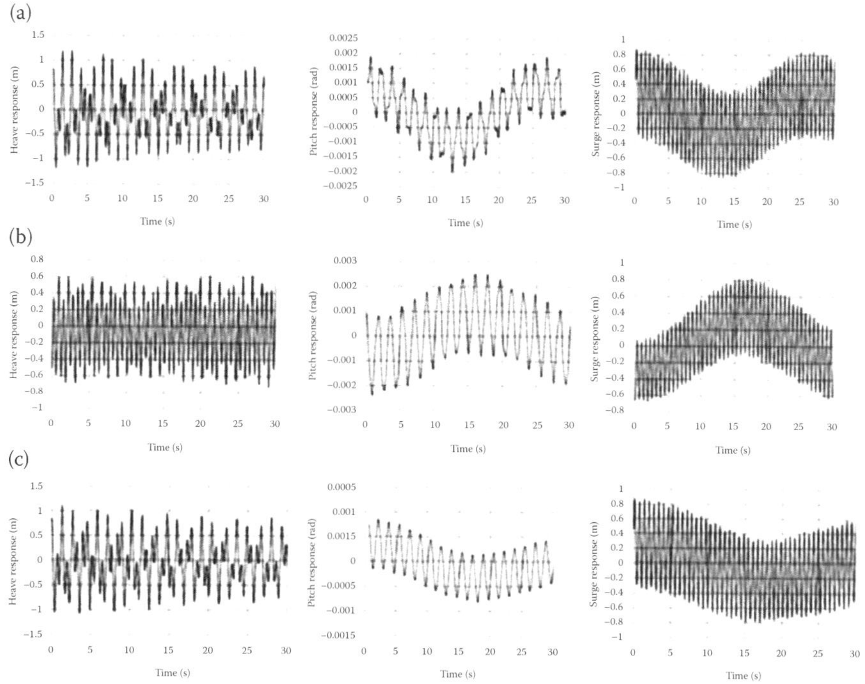

FIGURE B2.13 Response time histories of triangular TLPs under seismic forces and high sea waves. (a) Response of TLP_1, (b) response of TLP_2, and (c) response of TLP_3.

showing power spectral density (PSD) plots of response in three degrees of freedom; as well as (iii) figure showing the phase plots of the responses in order to ensure that the system is stable and periodic. PSD plots of the responses are obtained by direct Fourier transform of the response time histories. Prior to Fourier transform, data is smoothened by a data window using standard procedures suggested by Bringham (1974). Power spectral density curves are plotted for all example TLPs for a frequency range of 0–5 Hz whereas time histories are plotted for equivalent time period of 30 s. In general, it can be seen that all degrees of freedom show high-frequency response with no visible damping effects as the amplitudes do not die down with time.

Frequency responses are essentially narrow banded with energy concentrations as follows:

Heave responses show three distinct peaks namely (a) first at zero frequency (which is very small and could be neglected); (b) second at a mid-frequency, which is near the natural frequency of TLP_1 and TLP_2 and near twice the natural frequency of TLP_3. This peak is significantly smaller in TLP_1 compared to the others; and (c) third at 1.588 Hz, which is near the average of peak frequencies of PM and KT spectrums (PM: 0.07, KT: 2.5, average: $(0.07 + 2.5)/2 – 1.3$). The pitch responses also show three distinct peaks namely (a) first, near zero (0.04 Hz for TLP_1 and TLP_2 and 0.02 for TLP_3, respectively). This is significantly high for TLP_1 compared to the

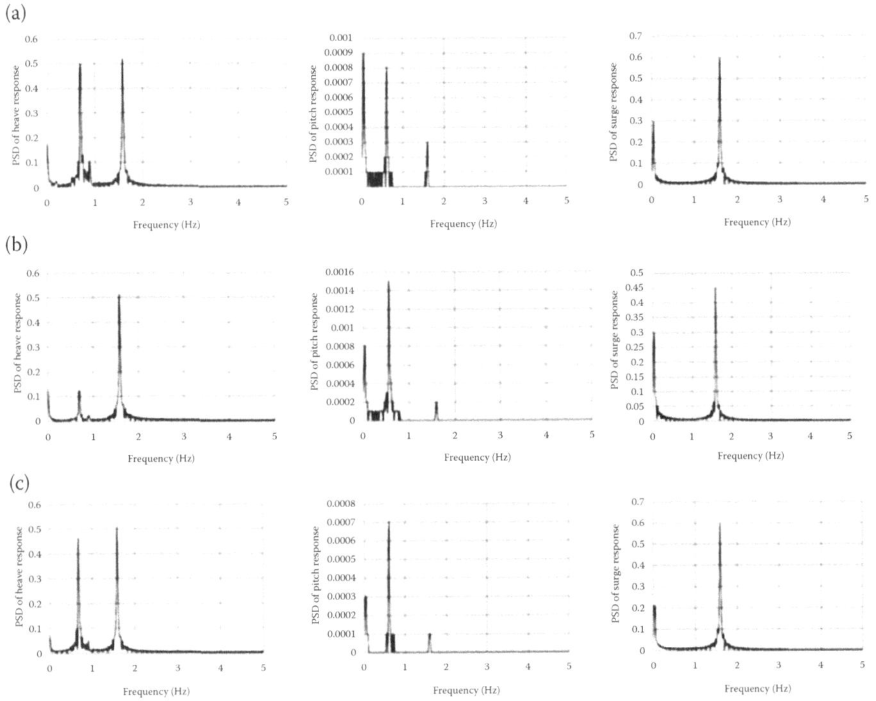

FIGURE B2.14 Frequency domain response of TLPs under seismic forces and high sea waves; PSD indicates power spectral density. (a) PSD plots of TLP_1, (b) PSD plots of TLP_2, and (c) PSD plots of TLP_3.

others; (b) second, near the natural frequencies for TLP_1 and TLP_2 and near twice the natural frequency for TLP_3; and (c) third, a little peak at 1.588 Hz. Clearly, the occurrence of peaks at 1.588 Hz is a significant manifestation of seismic excitations, which is influencing all the active degrees of freedom. Besides, TLP_1 and TLP_2 show a slightly broad-banded response (a lot of little peaks between the first and second peak). This could be seen as an advantage of deeper TLPs, because in reality, we have multiple frequency inputs to the system.

Therefore if the system is narrow-banded, it means that it will be excited by a lesser range of inputs and hence TLP_3 would be less susceptible to fatigue compared to the other cases. Surge responses show two distinct peaks namely (a) the first near zero (0.04 for TLP_1 and TLP_2, 0.02 for TLP_6) and also these peaks are almost at the same ordinates for TLP_1 and TLP_2 while being shorter for TLP_3; and (b) second at 1.588 Hz. The system response is negligible in the range of frequency higher than 2 Hz. Phase plots for all the cases presented, being elliptical in nature, affirm that all states are stable and periodic (but with a long period).

From the example problem, it is seen that the dynamic tether tension variation essentially caused due to earthquake motion is of a high order (about 65%). Clearly, the peaks seen in the response of all the three active degrees of freedom namely heave, pitch, and surge, occurring at the average sum frequencies of waves and

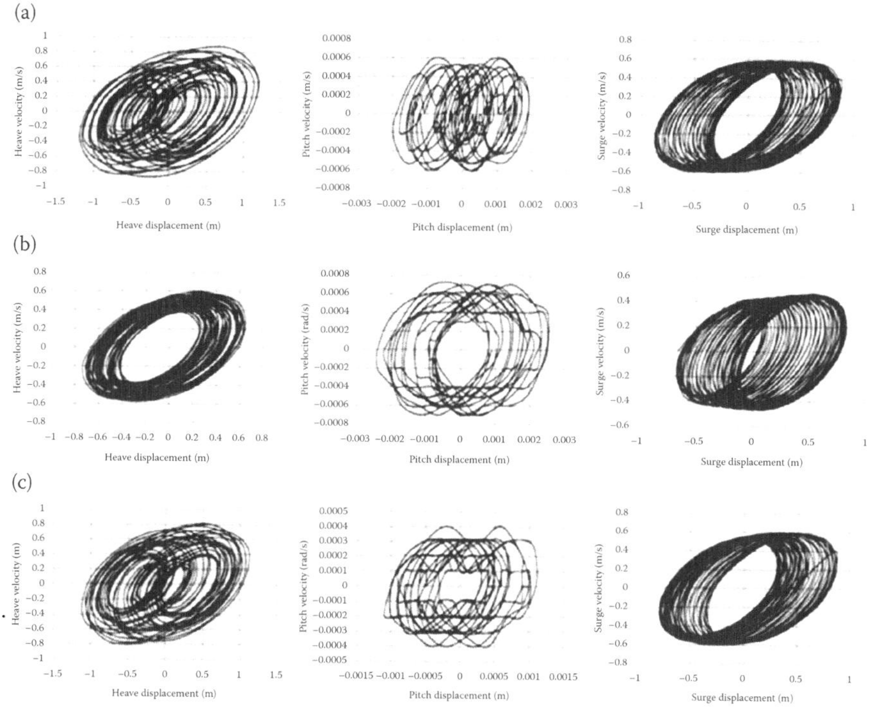

FIGURE B2.15 Phase plots of response of TLPs under seismic and high sea waves. (a) TLP1, (b) TLP2, and (c) TLP3.

earthquake ground motion is a significant influence of seismic excitations on TLP tethers under high sea waves. While TLPs are expected to be rigid in stiff degrees-of-freedom like heave, apart from affecting the integrity of the structure as a whole, the functionality of the platform might be lost if the displacements and corresponding velocities in the vertical directions are high. It is emphasized that TLPs should be analyzed for horizontal and vertical seismic excitations in the presence of distinctly high sea waves as this combination demands even stiff degrees of freedom like heave to be closely monitored for its response threshold values.

3 Reliability Analysis

3.1 INTRODUCTION

Environmental loads that act on offshore structures are not completely deterministic; only a few features of the loads are known (Ang and Tang, 1975). Some of the examples of such loading are extreme waves, seabed movement, wind loads on the superstructure, ice loads, impact loads caused by vessels, etc. (Srinivasan Chandrasekaran, 2015b). Apart from being random in nature, these loads also have a high probability of exceeding safe design limits, which can cause an overloading situation (Benjamin and Cornell, 1970). While safety is based on statistical judgment, reliability methods are based on engineering judgment. Hence, experience plays an important role in the accuracy of results of reliability methods. While applying reliability to offshore structures, which is one of class of structures of high economic importance, it is necessary to understand all the complexities that circumscribe the reliability analysis.

As seen from the earlier chapters, it is clear that a different class of uncertainties does not guarantee a deterministic approach for reliability analysis. These uncertainties clearly indicate that the reliability assessment of offshore structures should be considered from a probabilistic point of view. *Risk* is an extension of reliability, which also includes the consequences of failure. In engineering practice, one routinely encounters situations that involve some event, which has a particular probability of occurrence and if it occurs, has a specific set of consequences. With experience and engineering judgment, one should be able to assign a suitable probability to the occurrence of the event and some quantified magnitude or cost to the adversity associated with its occurrence (Chandrasekaran and Saha, 2011; Srinivasan Chandrasekaran, 2013; Cornel, 1969). This combination of uncertain event and adverse consequence is the determinant of risk. Alternatively, to express the condition of structure in a positive manner, reliability studies are conducted. As the term *risk* has a hidden meaning of chance of failure, reliability is expressed as (1-probability of failure); hence the reliability index is an indicator of safety and hence the term *reliability. Risk is an extension of reliability to address the consequences of failure; reliability terminates at the stage of assessing the probability of failure, while risk continues to address beyond this point.* It is important to know that risk conveys the financial component of unsafeness, which is required if one has to justify the cost investment toward risk mitigation or reduction; the latter is a common practice in oil and gas industries. Hence, risk covers a broad spectrum of the adverse effects on the following namely: (i) society; (ii) financial status; (iii) human life; and (iv) processing plant as a whole.

3.2 FUNDAMENTAL ANALYSIS

Two extreme types of structural elements that are commonly considered in system reliability analysis are namely: (i) brittle members; and (ii) ductile members. An element, which is brittle, becomes completely ineffective after it fails while a ductile element is able to maintain its load carrying capacity even after is fails.

3.2.1 SYSTEM WITH EQUALLY CORRELATED ELEMENTS

A special case involves a system in which the correlation coefficient is the same for all pairs of elements and its value is in the range [0,1]. Let us consider a system with n number of elements. Strength of the ith element ($i = 1, 2, 3,\ldots, n$) will be denoted as R_i. To estimate the probability of failure, the following assumptions are made: (i) strength of the elements are all normally distributed; (ii) strengths are all equally correlated; (iii) all applied loads are deterministic and time invariant; and (iv) all elements are designed to have the same reliability index β_e.

a. System in series with equally correlated elements

For n elements of a system in series, the probability of failure is given by (Thoft-Christensen and Dalsgard Sorensen, 1982)

$$P_f = 1 - \int_{-\alpha}^{\alpha} \left\{ \Phi\left(\frac{\beta_e + t\sqrt{\rho}}{\sqrt{1-\rho}} \right) \right\}^n \Phi(t)dt \tag{3.1}$$

where β_e is the reliability index or each elements, $\Phi()$ and $\Phi()$ are the standard normal cumulative distribution function (cdf) and probability density function (pdf), respectively, and ρ is the correlation coefficient.

System in parallel with equally correlated ductile elements Resistance of a parallel system with n ductile elements is given by

$$R = \sum_{i=1}^{n} R_i \tag{3.2}$$

where R is the system resistance and R_i is the resistance of each element. If we assume that resistance of all elements follow the same cdf, then the mean and variance of the system resistance can be expressed in terms of the element parameters μ_e and σ_e^2 as follows:

$$\mu_R = \sum_{i=1}^{n} \mu_{R_i} = n\mu_e \tag{3.3}$$

$$\sigma_R^2 = \sum_{i=1}^{n}\sum_{i=1}^{n}\rho_{ij}\sigma_{R_i}\sigma_{R_j}$$

$$= \sum_{i=1}^{n}\sigma_e^2 + \sum^{n}\sum_{i\neq j}^{n}\rho\sigma_e^2$$

$$= n\sigma_e^2 + \rho n(n-1)\sigma_e^2$$

$$= n\sigma_e^2(1-\rho+n\rho) \tag{3.4}$$

To determine the reliability index for the entire system, it is important to assess the relationship of β_e to the mean and standard deviation of each element. For the ith element, the limit state equation is given by

$$g(R_i) = R_i - q_i \tag{3.5}$$

where q_i is the load effect in the ith member for the given loading conditions. If strength of the ith element is normally distributed with mean μ_e and standard deviation σ_e, reliability index is given by

$$\beta = \frac{a_0 + \sum_{i=1}^{n} a_i\mu_{X_i}}{\sqrt{\sum_{i-1}^{n}(a_i\sigma_{X_i})^2}} \tag{3.6}$$

Reliability index for the element is given by

$$\beta_e = \frac{\mu_e - q_i}{\sigma_e} \tag{3.7}$$

By solving Equation 3.5 for q_i, we get

$$q_i = \mu_e - \beta_e\sigma_e \tag{3.8}$$

since μ_e, β_e, and σ_e are the same for all the elements, Equation 3.8 requires q_i to be the same for all elements. Hence,

$$\begin{aligned} q_{tot} &= nq_i \\ q_{tot} &= n\mu_e - n\beta_e\sigma_e \end{aligned} \tag{3.9}$$

Now the limit state equation for the entire system is given by

$$g(R) = R - q_{tot} \tag{3.10}$$

and the reliability index for the system is given by

$$\beta_{system} = \frac{\mu_R - q_{tot}}{\sigma_R} \tag{3.11}$$

Substituting, reliability index for a parallel system with equally correlated ductile elements is given by

$$\begin{aligned}
\beta_{system} &= \frac{n\mu_e - (n\mu_e - n\beta_e\sigma_e)}{\sqrt{n\sigma_e^2(1-\rho+n\rho)}} \\
&= \frac{n\beta_e\sigma_e}{\sigma_e\sqrt{n(1-\rho+n\rho)}} \\
&= \beta_e \frac{\sqrt{n^2}}{\sqrt{n(1-\rho+n\rho)}} \\
&= \beta_e \sqrt{\frac{n}{(1-\rho+n\rho)}}
\end{aligned} \tag{3.12}$$

3.2.2 System with Unequal Correlated Elements

In real structures, a system has elements that are unequally correlated. Following assumptions are made to do the reliability analysis of such systems: (i) strength of all the elements is normally distributed with identical distribution parameter μ_e and σ_e (ii) applied loads are deterministic and time invariant; and (iii) all elements are designed to have the same reliability index β_e

a. Parallel system with ductile elements

   Consider a parallel system with n perfectly ductile elements. The strength of the ith element ($i = 1, 2, \ldots, n$) will be denoted by R_i and strength of the entire system is denoted by R. The correlation matrix, describing the correlation between the elements is given by

$$[\rho] = \begin{bmatrix} 1 & \rho_{12}\cdots & \rho_{1n} \\ \rho_{21} & 1 & \rho_{2n} \\ \rho_{n1} & \rho_{n2}\cdots & 1 \end{bmatrix} \tag{3.13}$$

   The reliability index for the system is given by

$$\beta_{system} = \frac{\mu_R - q_{tot}}{\sigma_R} \tag{3.14}$$

$$\mu_R = n\mu_e \tag{3.15}$$

$$\sigma_R^2 = \sum_{i=1}^{n}\sum_{j=1}^{n}\rho_{ij}\sigma_{R_i}\sigma_{R_j}$$

$$= \sum_{i=1}^{n}\sigma_e^2 + \sum_{i}^{n}\sum_{i\neq j}^{n}\rho_{ij}\sigma_e^2$$

$$= n\sigma_e^2 + \sigma_e^2\left(\sum_{i}\sum_{i\neq 1}\rho_{ij}\right)$$

$$= \sigma_e^2\left[n + \sum_{i}^{n}\sum_{i\neq j}^{n}\rho_{ij}\right] \tag{3.16}$$

Using the above results and implying that all elements are subjected to the same load, system reliability is given by

$$\beta_{system} = \frac{n\mu_e - (n\mu_e - n\beta_e\sigma_e)}{\sqrt{\sigma_e^2\left[n + \sum_{i}^{n}\sum_{i\neq j}^{n}\rho_{ij}\right]}}$$

$$= \frac{n\beta_e\sigma_e}{\sqrt{\sigma_e^2\left[n + \sum_{i}^{n}\sum_{i\neq j}^{n}\rho_{ij}\right]}}$$

$$= \beta_e\sqrt{\frac{n^2}{n + \sum_{i}^{n}\sum_{i\neq j}^{n}\rho_{ij}}}$$

$$= \beta_e\sqrt{\frac{n}{1 + \frac{1}{n}\sum_{i}^{n}\sum_{i\neq j}^{n}\rho_{ij}}} \tag{3.17}$$

As the members are unequally correlated, an average correlation coefficient, $\bar{\rho}$, is defined as

$$\bar{\rho} = \frac{1}{n(n-1)}\sum^{n}\sum_{i\neq j}^{n}\rho_{ij} \tag{3.18}$$

Combining the above equations, we get system reliability as

$$\beta_{system} = \beta_e\sqrt{\frac{n}{1 + (n-1)\bar{\rho}}} \tag{3.19}$$

3.3 RELIABILITY BOUNDS FOR STRUCTURAL SYSTEMS

In system reliability, it is often convenient to work with Boolean variables. For example, consider n elements in the analysis for which it is assumed that each element can exist only in one of the two states namely: (i) failure; or (ii) non-failure. To describe these two states, Boolean variables, S_i and F_i are used and defined as follows for the ith element:

$$S_i = \begin{Bmatrix} 1 \text{ if the element is in a non-failure state} \\ 0 \quad \text{if the element is in a failure state} \end{Bmatrix} \tag{3.20}$$

$$F_i = 1 - S_i = \begin{Bmatrix} 0 & \text{if the element is in a non-failure state} \\ 1 & \text{if the element is in a failure state} \end{Bmatrix} \tag{3.21}$$

Based on the states of the elements, we can define the state of the system using Boolean variables as below:

$$\bar{S} = \{S_1, S_2, \ldots, S_n\} \tag{3.22}$$

$$\bar{F} = \{F_1, F_2, \ldots, F_n\} \tag{3.23}$$

System function is defined as $S_s(\bar{S})$ or $F_s(\bar{F})$ which indicates the state of the entire system. The function is defined as

$$S_s(\bar{S}) = \begin{Bmatrix} 1 & \text{if the element is in a non-failure state} \\ 0 & \text{if the element is in a failure state} \end{Bmatrix} \tag{3.24}$$

$$F_s(\bar{F}) = 1 - S_s(\bar{S}) = \begin{Bmatrix} 0 & \text{if the element is in a non-failure state} \\ 1 & \text{if the element is in a failure state} \end{Bmatrix} \tag{3.25}$$

For a series system, failure of one element means system failure. Therefore, the system function can be expressed as

$$S_s(\bar{s}) = S_1 S_2 \ldots S_n = \Pi_{i=1}^{n} S_i \tag{3.26}$$

If the ith element is in a failure state, then $S_i = 0$. Hence, the product will be equal to 0. If none of the elements has failed, then all of the S_i values are equal to 1 and hence the product is equal to 1. For a parallel system with perfectly ductile elements, if at least one element is in a non-failure state, then the system is in a non-failure state. Therefore, the corresponding system function can be written as

$$S_s(\bar{s}) = 1 - \Pi_{i=1}^{n} (1 - S_i) \tag{3.27}$$

Thus if one of the elements is in a non-failure state, then $(1 - S_i) = 0$ is true for that element; product in the above equation will be equal to 0 and the resulting function will be equal to 1. This is also true as the system is in a non-failure state. Using Boolean variables, one can calculate the expected values and probabilities. Consider the ith element in a system. If the state of the element is random, then the Boolean variable S_i is a discrete random variable. There are only two possible values [1 and 0] and there is a probability associated with each value as given by

$$E(S_i) = (1)\left[P\left(S_i = 1\right)\right] + (0)\left[P\left(S_i = 0\right)\right] = P\left(S_i = 1\right) \tag{3.28}$$

Note that $P(S_i = 1)$ is the same as $P(F_i = 0)$. Also note that $P(F_i = 0) + P(F_i = 1) = 1$ as the sum of all the probabilities over the entire range of possible discrete values must be equal to unity. Substituting these results in Equation 3.28, we get

$$E\left(S_i\right) = P\left(S_i = 1\right) = 1 - \left[P\left(F_i = 1\right)\right] \tag{3.29}$$

Similarly, one can determine the expected value of F_i as follows:

$$E(F) = (1)\left[P\left(F_i = 1\right)\right] + (0)\left[P\left(F_i = 0\right)\right] = P\left(F_i = 1\right) \tag{3.30}$$

Probability of failure of the entire system, P_f, can be found using the same approach:

$$\begin{aligned} E\left[F_s\left(\bar{F}\right)\right] &= (1)P\left[F_s\left(\bar{F}\right) = 1\right] + (0)P\left[F_s\left(\bar{F}\right)\right] = 0 \\ &= P\left[F_s\left(\bar{F}\right) = 1\right] = P_f \end{aligned} \tag{3.31}$$

3.4 APPLICATION OF STRUCTURAL CODES ON SAFETY

The intent of a design code is to provide a minimum safety level and to ensure that the structure intends to serve up to its design life without any failure (Srinivasan Chandrasekaran and Bhattacharyya, 2014; Srinivasan Chandrasekaran, 2014a). Most of the international codes that are currently in practice use deterministic formulas. As a designer, one can classify the design codes into four levels, based on approach to reliability: (i) level I codes use deterministic design formulas. The safety margin is introduced through central safety factors; (ii) level II codes define the design acceptance criterion in terms of the closeness of the actual reliability to that of the target reliability index; (iii) level III codes demand a complete reliability analysis to quantify the probability of failure of the structure under various loading states; and (iv) level IV codes use the total expected cost of the design as the optimization criterion.

3.5 LIMIT STATE FUNCTIONS

In structural reliability analysis, three types of limit states are considered:

1. *Ultimate limit state* (*ULS*): Ultimate limit states (ULSs) are mostly related to the loss of load-carrying capacity.
2. *Serviceability limit states* (*SLS*): SLS are related to gradual deterioration, maintenance costs.
3. *Fatigue limit states* (*FLS*): FLS are related to loss of strength under repeated loads.

Each limit state function is associated with a particular limit state. In the following section, let us discuss the modeling of limit state function and its associated characteristics.

3.6 CHARACTERISTIC VALUE OF BASIC VARIABLES

3.6.1 Treatment of Geometric Variables

Let q be the total load and R be the resistance. Limit state function is given by

$$g(R,Q) = R - Q \tag{3.32}$$

3.6.2 Treatment of Material Properties

In simple terms, moment-carrying capacity of a section is $R = F_y\, Z$ where F_y is the yield stress and Z is the plastic section modulus. Substituting in the above equation, limit state function, we get

$$g(F_y \cdot Z, Q) = F_y Z - Q \tag{3.33}$$

3.6.3 Treatment of Load and Other Actions

Considering that the member is subjected to combination of loads, total demand or load effect on the member consists of contributions from dead load (D), live load (L), wind load (W), and earthquake load (E). If $Q = D + L + W + E$, then limit state function is given by

$$g(F_y, Z, D, L, W, E) = F_y Z - D - L - W - E \tag{3.34}$$

On the other hand, limit state function can be a function of many variables, load components, influence factors, resistance parameters, material properties, dimensions, analysis factors, and so on.

3.6.4 Evaluations of Partial Coefficient

The basic form of the LRFD equation is given by

$$\varnothing R \geq \sum \gamma_i Q_i \tag{3.35}$$

Design equation contains partial safety factors ϕ and γ, which must be calibrated based on the target reliability index, adopted by the code. However, due to inherent complexities leading various types of uncertainties, both in load and resistance, make limit state function more complex.

3.7 STOCHASTIC MODELING

Modeling of elementary variable implies that we have some information, however inadequate it may be, as shown in Figure 3.1. Engineering judgment is used to characterize the variable based on limited experience.

In a given interval, a degree of possibility is associated with each value of the variable. For instance, V ranges between $[V_{min}, V_{max}]$ with each value of possibility associated with V. Depending on the approach to define safety, one typically uses characteristic values obtained generally by increasing a typical value. In more appropriate terms, it is better to represent this variable by a known distribution with density $f_v(_v)$. To include the random nature of variable, it is denoted by $V(\omega)$, where $\omega \in \Omega$, is the event space. Reliability approach therefore should be compatible with the available information. It is better to use a stochastic model to represent the random variable as they are inclusive of representing variability with an appropriate pdf. Two main approaches are followed namely: (i) naturalist's approach and (ii) physicist's approach.

The naturalist's approach relies on the observation of a sample, which estimates mean, variance, and other statistical moments. Statistical methods offer estimates of these random variables and based on the best adjustment, an appropriate pdf is determined. The chosen function is verified with "goodness of fit" test, Chi-square test, or Kolmogorov–Smirnov tests. The physicist's approach seeks to understand the variability of material behavior on a macroscopic scale based on the state of the material. For example, Weibul's model shall be used to identify the pdf, which includes the mechanism of failure of material, containing internal defects of random dimensions. Most of the geometric-based and material-based uncertainties can be included in the model but the result of reliability calculations depends on the quality of the data, which is generally insufficient. This may be mainly due to the size of the test samples; the larger

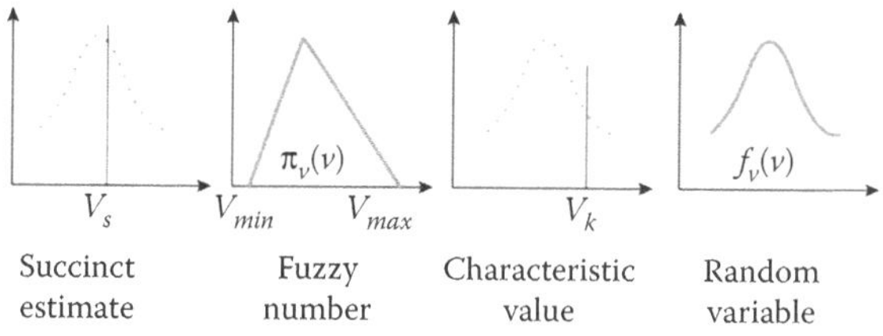

FIGURE 3.1 Representations of variables.

the number of tests, the greater the chances of stumbling upon an unsatisfactory test. This paradox is resolved by the probabilistic approach, which shows a better accuracy of information. For instance, Gaussian distribution is one of the most common probability distributions of the random variable, but represents an infinite domain, which is unacceptable in reliability analysis. Nevertheless, if truncations are made, Gaussian distribution is certainly a very well adapted distribution as it represents central tendencies satisfactorily. Typical Gaussian density function is shown in Figure 3.2.

It is interesting to note that for rare events, the reliability calculation concerns the distribution tails for which information is evidently rare. In such a situation, the approach using extreme value statistics provides a preliminary response and a physical examination. It is possible to compare different assumptions to search for an envelope, in such cases.

Considering a function $V(x, t, \omega)$ on space, the time and the random set $\omega = \Omega$, when fixing the random set element to $V(x, t, \omega_0) = V(x, t, \omega_0)$, the outcome is a trajectory of this function for a fixed element of the random set. The term stochastic process (or random process) denotes such a function. It can be indexed only by time or in space; for example, peak ground acceleration of an earthquake signal is a spatial distribution of an elasticity modulus. In such cases, it is necessary to perform discretization, as shown below:

$$V(x,t,\omega) \rightarrow \{V(t,\omega)\} = \begin{Bmatrix} V_1(x = x_1, t, \omega) \\ V_2(x = x_2, t, \omega) \\ \vdots \\ V_n(x = x_n, t, \omega) \end{Bmatrix}$$

$$[\rho] = \begin{bmatrix} 1 \\ \vdots \end{bmatrix} \tag{3.36}$$

$$[\rho] = \begin{bmatrix} 1 & \rho_{12} & \rho_{1n} \\ \rho_{21} & 1 & \rho_{2n} \\ p_{n1} & \rho_{n2} & 1 \end{bmatrix}$$

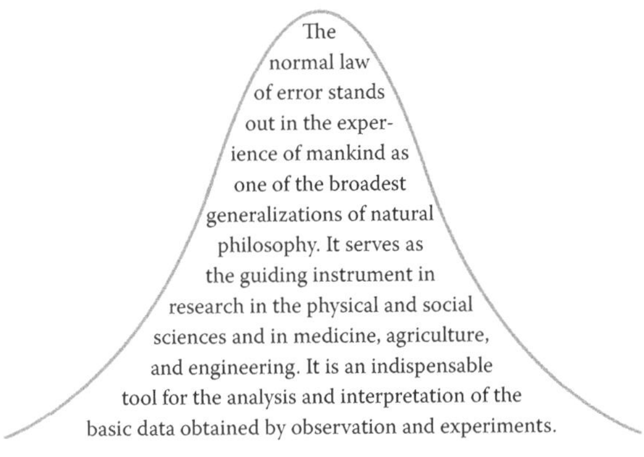

FIGURE 3.2 Gaussian density plot.

Such a discretization quickly results in large vectors, which can become mathematically complex to handle; most of the current statistical tools and software have limitations in handling such problems. Alternatively, it is common practice to rely on stochastic modeling, indexed by time. However, in many cases it is possible to use an extreme value distribution to express the maximum and minimum for a given period. For example, probability of maximum wind speed in year not exceeding a limiting quantity can be expressed as

$$F_v(v) = Prob(V \leq v) \tag{3.37}$$

In the N successive year, it is possible to consider that annual probabilities are independent, with the same distribution functions. Modeling the maximum speed for N years is given by

$$Prob(V_1 \leq v)\,Prob(V_2 \leq v) = F_{Y_N}(v) \tag{3.38}$$

The study of extreme values, when $N \rightarrow \alpha$ has led to different extreme value distributions namely: Gumbel, Freshet, and Weibull distributions. It is thus possible to represent a repeated function of time by random variables, which is an essential step in the reliability analysis.

3.8 MECHANICAL MODELING

The purpose of modeling is to construct explanatory and predictive models of physical phenomena. Theoretical models and the corresponding numerical solutions are shown in Figure 3.3. There will be always a decent difference between behavior of the model of the structure and that of the physical reality. However, modeling deviation is normal and acceptable; procedures for validation of models are used to control it. It can be seen that it is not totally random but contains a systematic bias. On the other hand, gross errors must be prevented by a process of quality assurance.

Models are evaluated to know how well they are represented with that of the physical phenomenon. Mathematical models should explain and represent the physics as close as possible while numerical modeling must control the accuracy of the results so that deviations between the physics and its image must be the least. It is necessary to have a validated mechanical modeling approach. Validation of the model and the numerical solution (density of the finite element mesh, convergence tests, etc.) have a measure of the bias of the model and of a deviation random variable. The next stage is to check whether uncertainties in the data influence the variable of the model. Data introduced in the numerical model are known with only a few significant digits (very often not more than two or three). They are uncertain and the designer must forgo the certainty on the significance of the values that he chooses when he initiates a calculation code. Mechanical models are used to evaluate the expected internal strengths or stresses. Required resources, which are the availability of material and their resistance, are shown in Figure 3.4. Theoretical analysis is based on a failure scenario, combining the evaluation of internal strength and that of their resistance,

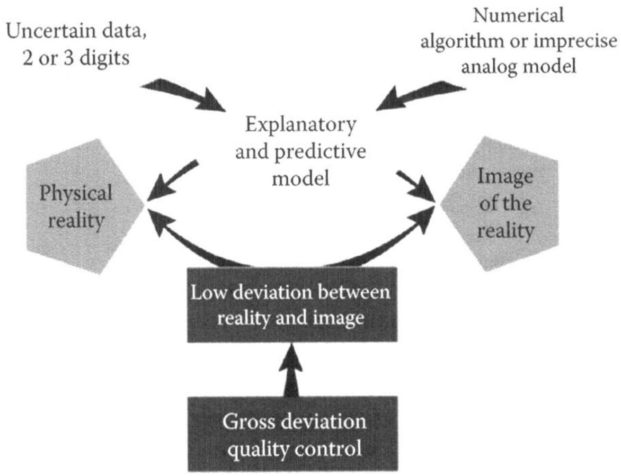

FIGURE 3.3 Approximated representation of physical reality.

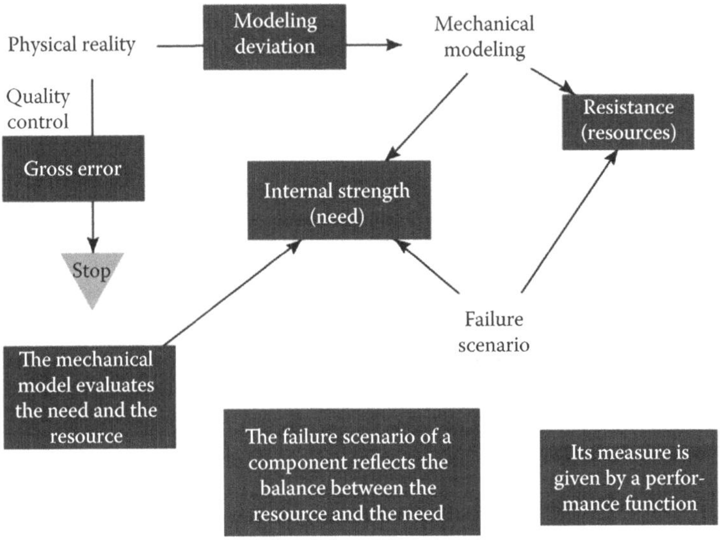

FIGURE 3.4 Failure scenarios in mechanics.

which are a function of the elementary variables. Failure scenario reflects the balance between the resource and the need and is expressed by a performance function.

Mechanical models should be able to estimate the basic parameters like mean, standard deviation, etc. It should also have the compatibility to adjust a distribution of internal strength or resistance variables, if required. Furthermore, it should also be able to estimate the probabilities of failure scenarios of occurrences; theoretical probabilities should be conditioned by the practical reliability approach. Therefore, an efficient reliability analysis should provide decent reliability sensitivity. This can be achieved by applying a coupling between mechanical and stochastic models.

3.9 MECHANICAL MODEL AND RELIABILITY COUPLING

The mechanical model ensures the transition between input data and the output variables as shown in Figure 3.5. Computation of the statistical parameters of output variables, with respect to that of the input data is called reliability sensitivity analysis. Deterministic sensitivity analysis consists of computing the gradient around a point. Reliability sensitivity analysis searches for a relationship between the respective coefficients of variation (COV) of the output to input variable.

Search for the statistical parameters of the response of a mechanical model, as a function of the variability of input data, around a known value is the actual complexity of the reliability analysis. Two methods are commonly used namely: the Monte Carlo method and Perturbation method. The Monte Carlo method proceeds by simulations. It constructs a sample from which the statistical moments can be deduced without any *a priori* limitation of order. Perturbation requires calculation of derivatives of the performance function with respect to the random data. It is generally limited to the first two moments. With regard to the most probable failure point, which is very important in reliability analysis, both the methods differ. In Monte Carlo simulation, the sensitivity analysis relies on the capacity to construct the synthetic data samples. It depends on the quality of the random number generator and on the capacity to identify samples of output variables or processes. Reliability analysis requires an additional scenario as shown in Figure 3.6. It separates the situation that the designer decides to consider acceptable from those that are unacceptable. Often it is possible to construct two independent numerical models: one for the internal strength and other for the evaluation of resistances. In such cases, the difference between the internal strength and resistance is a margin, which is the random variable. In the statistical context, the objective of the reliability analysis is therefore to evaluate the probability that this margin has a positive value.

Based on the input data related to loads, limit state, material resistance characteristics, and suitability of appropriate reliability model, the objective of the reliability

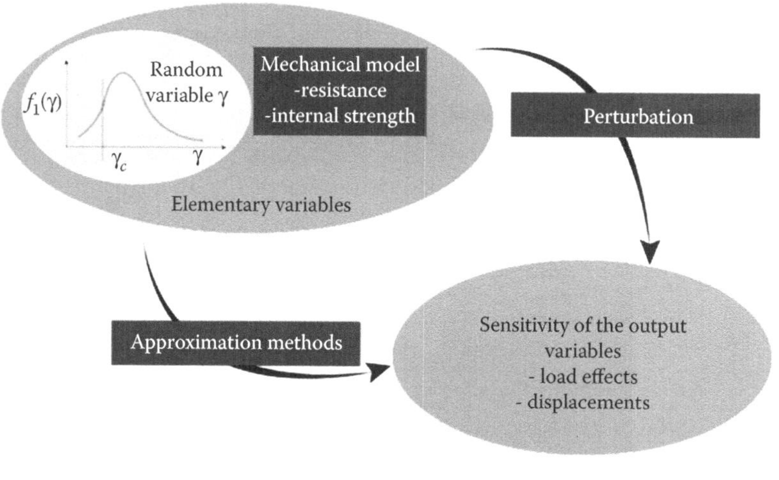

FIGURE 3.5 Reliability sensitivity analyses.

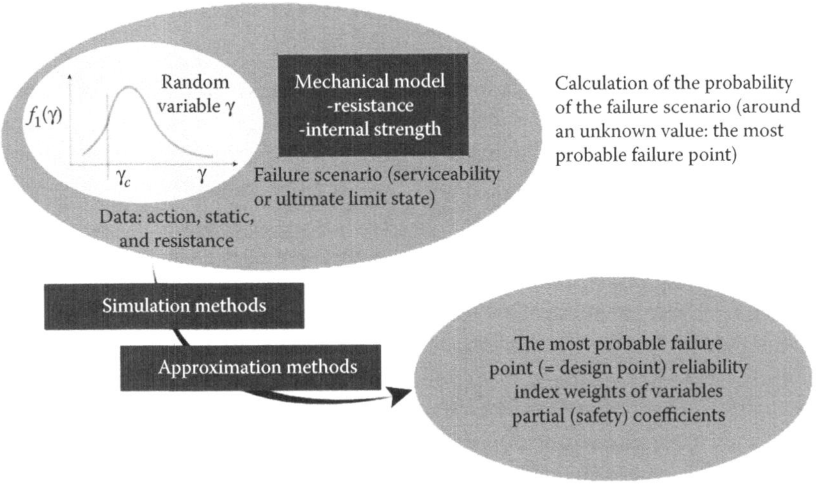

FIGURE 3.6 Reliability analysis.

analysis is to calculate a reliability index from approximating the probability of failure. It is also important to note that at the probable failure points (also called a design point), the sensitivity factor to failure and evaluation of partial (safety) coefficients are also considered. Therefore, one can infer that the numerical procedure of the reliability model excites the internal strength and resistance by engineering judgment, applied to the available data to produce the required results. The Monte Carlo simulation methods or other approximation methods are some of these excitations strategies.

3.10 COMPLEXITY OF MECHANICAL MODEL AND RELIABILITY COUPLING

Four criteria, depending on the modalities they take, help define the complexity of the coupling of mechanical model and reliability as shown in Figure 3.7.

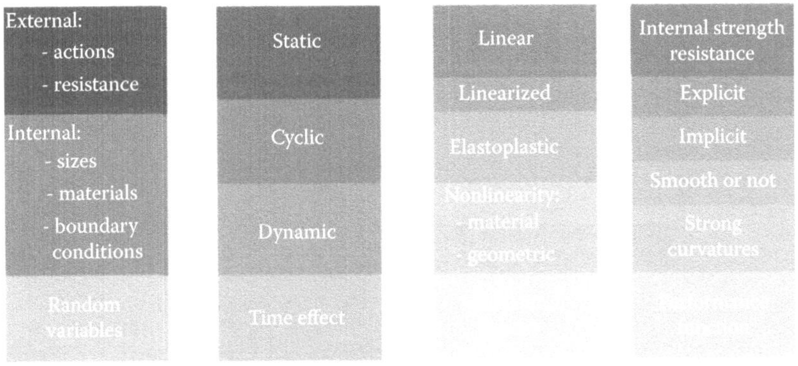

FIGURE 3.7 Coupling of mechanical model and reliability coupling.

Complexities arise in coupling the mechanical model to the reliability analysis from many factors namely: (i) time-variant properties of loads and materials; (ii) degree of approximation used in the level of analyses; (iii) inclusion of secondary effects that arise from geometric and material nonlinearity; and (iv) choice of limit state functions and appropriate probabilistic models. It is rather difficult to include material degradation that arises from aging in the mechanical model; but it is unfortunate to note that this has a significant influence on the reliability estimates. Also, a few limitations exist in the mechanical modeling of form-dominated offshore structures, which cannot be examined without approximations. For example, see Section 3.10.1.

3.10.1 Complexities in Geometric Modeling of Hinged Joints

A new geometric form of offshore platform namely the offshore triceratops is assessed for its suitability for deep-water oil exploration. It consists of a deck, which is supported by buoyant leg structures (BLS) through ball joints. The reduction in response shall be attributed to the presence of a hinged joint between the deck and BLS, which restrains transfer of rotational displacements from the BLS to the deck and vice versa. A scaled model of the platform is designed; mass properties of this scaled model are derived on the basis of the conceptual idea given in the literature (White et al., 2005). A scale ratio of 1:72 is selected on the basis of the material available to map the equivalent mass distribution, as desired. Figure 3.8 shows the model considered for the study. Payload is equally distributed to each BLS unit; this equal distribution helps to maintain the desirable longitudinal and lateral centers of gravity. Payload includes the mass of different components namely: (i) the platform; (ii) drilling rig; (iii) riser tension; (iv) cranes; (v) storage facilities; (vi) mud pit; (vii) generator; (viii) fire-fighting equipment; (ix) blow out operator; and (x) living quarters including the life boats, helideck with helicopter mass, pumps, and pipes. The mass of each of the above are computed and used in this study (Chakrabarti, 2005). Generally, the drill rig will be placed at center of the deck to have similar mass distribution. But in the present configuration, drill rig is placed on top of one of the BLS and risers are planned to connect at the moon pool of the BLS unit; this makes effective usage of the deck area. Figure 3.9 shows a typical hinged joint that connects BLS with the deck.

It is now important to estimate the stiffness of the hinged joint under the combined action of axial force and moment, in the installed condition. It is because restraints offered by the hinged joint significantly influence the response of the deck in comparison to that of the BLS units under wave loads. This attempt shall be made only during the installed condition of the platform because joint reactions cannot be artificially incorporated in any testing facility in the joint to assess their stiffness. Joint stiffness and damping are estimated by applying the known force near the joint. This becomes one of the serious limitations in coupling the mechanical modeling with reliability analysis, because the resistance variables considered for the hinged joint will significantly influence the failure of the platform. However, this issue is resolved in the experimental investigations. The applied force is measured using a ring-type load cell. The rotational response of the BLS unit is recorded using the inclinometer. Moment applied per unit rotation is considered as

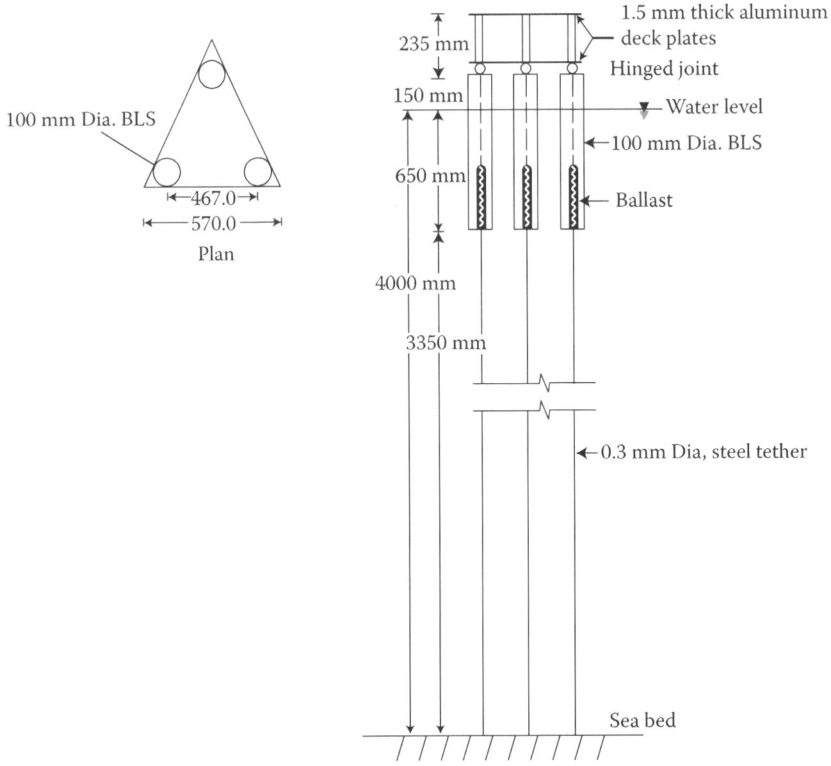

FIGURE 3.8 Offshore triceratops.

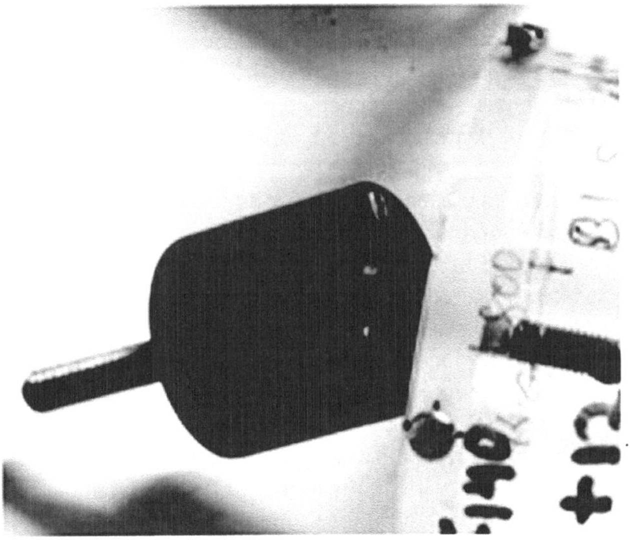

FIGURE 3.9 Hinged joint connecting deck and buoyant legs.

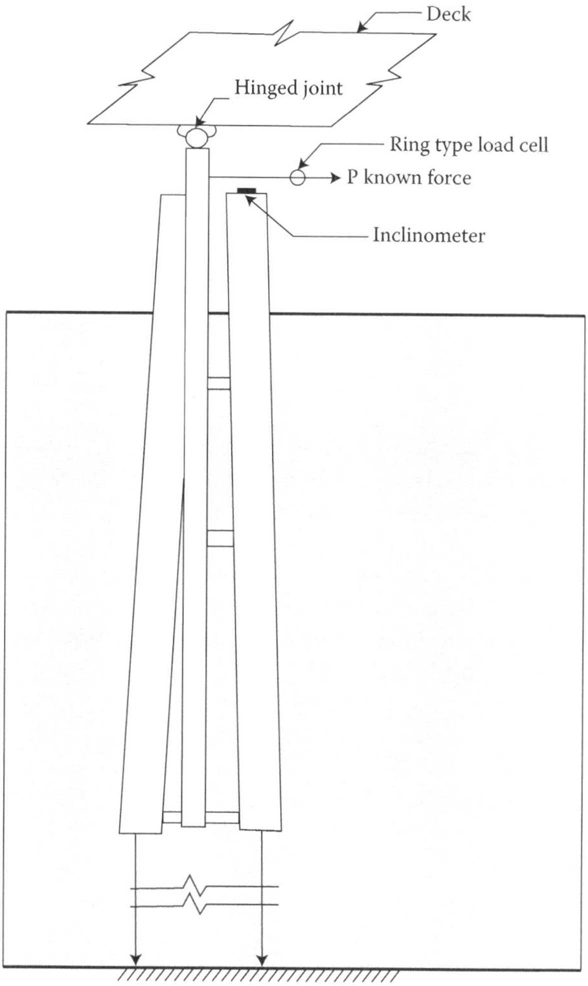

FIGURE 3.10 Schematic diagram to find stiffness and damping of the hinged joint.

stiffness of the hinged joint. The damping of the joint is estimated using logarithmic decrement method from the free oscillations of the BLS. Schematic diagram of the experimental set up for estimating stiffness of the hinged joint is shown in Figure 3.10. The above exercise is also done in pitch and roll degrees-of-freedom to obtain moment–curvature relationship of the hinged joint in pitch and roll degrees-of-freedom, as shown in Figure 3.11.

3.11 STOCHASTIC PROCESS

A stochastic process is an indexed set $\{X[t], t \in T\}$ of random variables $X[t]$, where all $X(t)$ are defined on the same sample space Ω. Two different kinds of variables are

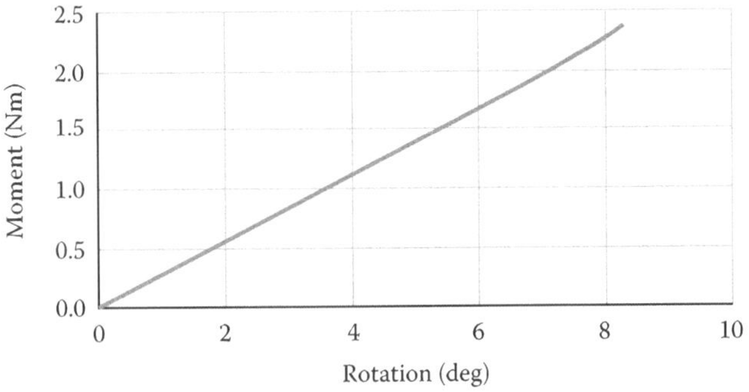

FIGURE 3.11 Rotational stiffness of hinged joint in rotational degrees of freedom (scaled values).

involved in the process namely: (i) stochastic variables $X(t)$; and (ii) variable t, which is called as index. The index set T is typically a time interval, but can be any kind of finite set. Probabilistic structure of a stochastic process is described in a way similar to that of random vectors. If the index set is a finite set, then the stochastic process forms a random vector. The fact that a stochastic process is a set of random variables makes it natural to describe its probabilistic structure in a way similar to random vectors, but in this case, the index set is infinite.

For a fixed sample space Ω the outcome of the set of stochastic variables form an ordinary function called a realization. If the experiment is repeated, the new realizations will not be the same as in Figure 3.12, but its probabilistic contents will be the same. In Figure 3.13, values of $X(t_1)$ are shown for a number of realizations and also the associated density function $f_{\{X\}}(x{:}t_1)$. Given two instants of time t_1 and t_2, some correlation between $x(t_1)$ and $x(t_2)$ exist, especially when the time difference $|t_1 - t_2|$

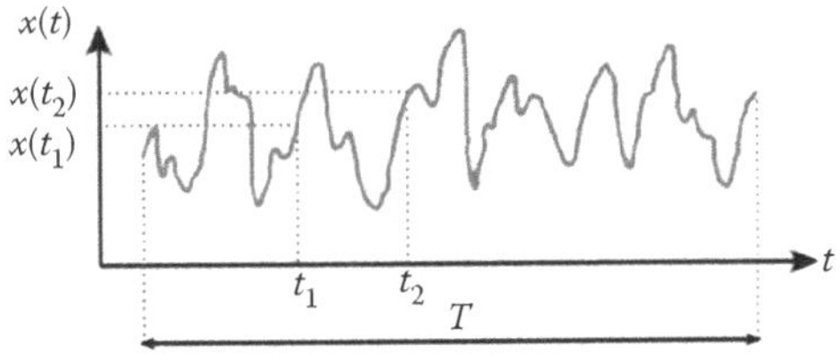

FIGURE 3.12 Realization of the stochastic variable.

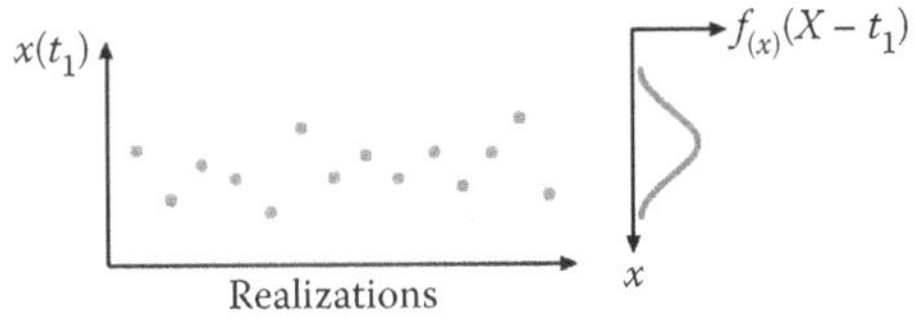

FIGURE 3.13 Realization and the corresponding density function.

is small. This is taken into consideration through the joint distribution function $F_{\{X\}}(x_1, x_2{:}t_1, t_2)$, which is defined by

$$F_{\{X\}}(x_1,\ x_2 : t_1,t_2) = P\Big(\big(X(t_1) \le x_1\big) \cap \big(X\big)(t_2) \le x_2\Big) \tag{3.39}$$

This joint distribution function for arbitrary $(t_1, t_2) \in T^2$ is called the joint distribution function of order 2. The corresponding joint density function of order 2 is given by

$$F_{\{X\}}\left(x_1,\ x_2 : t_1,t_2\right) = \frac{\partial^2 F\{X\}(x_1,\ x_2\ :\ t_1,t_2)}{\partial X_1 \partial X_2} \tag{3.40}$$

In describing a stochastic process, following functions (of time) are important. The mean value function $\mu_x(t)$ is defined as the expected value of $X(t)$ as

$$\mu_x(t) = E[X(t)] = \int_{-\infty}^{\infty} x f_{\{X\}}(x;t)dx \tag{3.41}$$

The autocorrelation function $R_{xx}(t_1, t_2)$ is equal to the following joint moment of the random variables $X(t_1)$ and $X(t_2)$ and is given by

$$R_{xx}(t_1,t_2) = E\left[X(t_1)X(t_2)\right] = \int_{-\infty}^{\infty}\int_{-\infty}^{\infty} x_1 x_2 f_{\{X\}}(x_1,x_2;t_1,t_2)dx_1 dx_2 \tag{3.42}$$

The autocovariance function $C_{xx}(t_1, t_2)$ is the covariance of the random variables $X(t_1)$ and $X(t_2)$ and is given by

$$C_{xx}(t_1,t_2) = E\Big[\big(X(t_1) - \mu_x(t_1)\big)X(t_1) - \mu_x(t_2)\Big] = R_{xx}(t_1,t_2) - \mu_x(t_1)\mu_x(t_2) \tag{3.43}$$

By setting $t_1 = t_2 = t$, variance function $\sigma_X^2(t)$ of the random variables $X(t)$ is obtained as

$$\sigma_x^2(t) = C_{xx}(t,t) = R_{xx}(t,t) - \mu_X^2(t) \tag{3.44}$$

Autocorrelation coefficient $\rho_{xx}(t_1, t_2)$ is defined by

$$\rho_{xx}(t_1,t_2) = \frac{C_{xx}(t_1,t_2)}{\sigma_X(t_1)\sigma_X(t_2)} \tag{3.45}$$

For an important group of stochastic processes, all finite dimensional distributions are invariant to a linear translation of the index set. Such processes are called strictly homogeneous or when the index parameter is time, strictly stationary. When this invariant assumption only holds for distributions of order 1 and 2, then the process is called weakly homogeneous or weakly stationary. An important consequence of the assumption of stationary is that $f_{\{X\}}(x;\,t)$ and $F_{\{X\}}(x;\,t)$ becomes independent of t so that we can omit reference to t. Further, the second-order distributions will only depend on the difference of the index parameter τ $(=t_1 - t_2)$. The same is true for all other statistics above. In practical applications, modeling of a physical quantity by a stochastic process is based on a single realization of a stationary process. If only one realization is at hand, then the mean value is given by

$$\mu = \frac{1}{T}\int_0^T x(\tau)d\tau \tag{3.46}$$

If this time average approaches μ_X for $T->\infty$, then the process is ergodic in the mean value. In the same manner, a process is ergodic in correlation, if the following condition is satisfied:

$$R(\tau) = \frac{1}{T-\tau}\int_0^{T-\tau} x(t+\tau)x(t)dt \tag{3.47}$$

If the above value approaches $R_{xx}(\tau)$ for $T->\infty$, then the process is ergodic. Note that stationary property is an assumption behind the definition of an ergodic process so that any ergodic process is stationary but not vice versa.

3.12 GAUSSIAN PROCESS

A process $\{X(t),\ t\epsilon T\}$ is Gaussian if the random variables $X(t_1)$, $X(t_2)$,…,$X(t_n)$ are jointly normal for any n, t_1, t_2,…,t_n.

$$f_{\{X\}}(x_1,\ldots,x_n;t_1,\ldots,t_n) = \frac{1}{(2\pi)^{n/2}\,|\overline{\overline{C}}^{1/2}}\,e^{-1/2\sum_{i,j=1}^{n}(x_i-\mu_X(t_i)M_{ij}(x_j-\mu_X(t_j)))} \tag{3.48}$$

where $\overline{\overline{C}}$ is the auto covariance matrix and is given by

$$\overline{\overline{C}} = \begin{matrix} C_{xx}(t_1,t_1) & C_{xx}(t_1,t_2) & \ldots C_{xx}(t_1,t_n) \\ \vdots & \vdots & \vdots \\ C_{xx}(t_n,t_1) & C_{xx}(t_n,t_2) & \ldots C_{xx}(t_n,t_n) \end{matrix} \tag{3.49}$$

where, M_{ij} is the (i, jth) element in C^{-1}. It is clear from the definition that a Gaussian process is completely determined by the mean value function $\mu_X(t)$ and the auto covariance function $C_{xx}(t_1,\,t_2)$. Therefore, a stationary Gaussian process is always strictly stationary.

3.13 BARRIER CROSSING

For a stochastic process, it is also important to know the number of crossings of a given barrier, which can be helpful to obtain the Dirac-delta function, indirectly. Figure 3.10 shows a realization $x(t)$ in the interval $[t_1;t_2]$ of a stochastic process $\{X(t)\}$ and a constant barrier $x(t) = \xi$. The number of up crossings of this barrier in the time interval $[t_1;t_2]$ is four, which are marked in Figure 3.14.

To solve the problem of estimating the expected number of positive passages of a given barrier, Heaviside step function is useful, which is given by

$$H(x) = \begin{cases} 0 & \text{for } x(t) < 0 \\ \frac{1}{2} & \text{for } x(t) = 0 \\ 1 & \text{for } x(t) > 0 \end{cases} \tag{3.50}$$

By formal differentiation of the function, one gets the Dirac-delta function $\delta(x)$, wherein only a definite set of values can be assigned to (x) is given by

$$\delta(x) = \lim_{\epsilon \to 0} \frac{1}{\sqrt{2\pi}\epsilon} e^{-x^2/2\epsilon^2} \tag{3.51}$$

3.14 PEAK DISTRIBUTION

The results derived from the barrier crossing can be used to investigate the statistics of the peak distribution of a stochastic process $\{\dot{X}(t)\}$, as peaks or troughs occur when the stochastic process $\{X(t)\}$ has a zero crossing. Interestingly, number of zero crossings of $\{\dot{X}(t)\}$ is equal to the number of extremes in $\{X(t)\}$. For $\{X(t)\}$ a narrowband Gaussian process, distribution of the peaks can be determined in a simple manner. Expected number of peaks above the level $\xi(\xi > 0)$ per unit time is equal to the expected rate of crossings of the barrier ξ, that is, equal to $E[N_+'(\xi)]$. Similarly, the expected total number of peaks per unit time is equal to the expected rate of zero

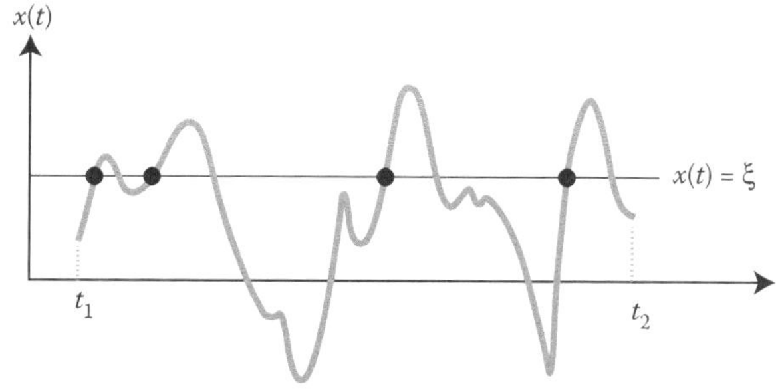

FIGURE 3.14 Realization of stochastic process.

crossings $E[N'_+(0)]$. Therefore, the expected relative number of peaks above ξ per unit time is given by

$$\frac{E[N'_+(\xi)]}{E[N'_+(0)]} = e^{-\xi^2/2\sigma_x^2}. \tag{3.52}$$

The distribution function $F_{\equiv}(\xi)$ for the peak magnitude ($\xi > 0$) is as follows:

$$F_{\equiv}(\xi) = 1 - e^{-\xi^2/2\sigma_x^2}, \quad 0 \le \xi < \infty \tag{3.53}$$

and the density function $f_{\equiv}(\xi)$ by

$$f_{\equiv}(\xi) = \frac{\xi}{\sigma_x^2} e^{-\xi^2/2\sigma_x^2}, \quad 0 \le \xi < \infty \tag{3.54}$$

This will follow Rayleigh distribution.

3.15 FATIGUE RELIABILITY

3.15.1 Discrete Wave and Spectral Methods of Fatigue Analysis

Fatigue is an important consideration in the design of offshore structures, especially when the connections are welded as in steel construction (ABS, 2003; 2004). Different parameters like variability of sea state, complexities involved in modeling the hydrodynamic loading and structural behavior (especially the dynamic response), stress concentration in tubular joints and difficulties associated with models for fatigue damage are vital in estimating the reliability of offshore structures (Aruliah, 1976; AWS, 1972; Baltrop and Adams, 1991). Fatigue damage can be assessed by two methods namely: (i) discrete wave approach and (ii) spectral approach. The discrete wave approach consists of simplified probabilistic models for the wave environment. This is a simplified approach, which uses a recommended set of waves of specified amplitude, frequency, and number or directly a stress range distribution (Ding et al., 2005; Ferry Borges and Castanheta, 1971). Spectral approach employs a random process model for the short-term sea state, an array of which constitutes a long-term wave environment, which is typically represented in the form of a wave scatter diagram/bi-variant histogram.

The predominant loading on offshore structures being wave loading, cyclic stresses induce fatigue damage, especially at the tubular joints of steel jackets and towers, which have high stress concentration zones (Ganapathy, 1996; Goda, 1970). The fatigue damage is aggravated by possible dynamic amplifications of the structural response as well. Fatigue assessment of an offshore structural element consists of establishing the fatigue demand of that element and comparing it with its fatigue strength (Goodwin et al., 1999; Hasan et al., 2009). Considering the stress range, which denotes the absolute sum of a cyclic stress history about a mean stress, fatigue demand is quantified in terms of the stress ranges experienced by a structural element due to various loads encountered by the platform. Basic fatigue data that is of relevance is the *S–N* curve, which represents the stress range and cycles-to-failure curve; such curves are generated using appropriate experimental data (Haver et al., 2001; Hovde et al., 1997).

The primary source of loading in offshore structures arise from (wind generated) ocean waves, which is a cyclic loading (Kanegaonker et al., 1982; Larsen and Mathisen, 1996). This may be described in two ways namely: (i) either using a discrete wave approach or (ii) using a stochastic model. While the former models nondeterministic nature of wave heights, the latter represents the sea state in spectral character, which is more realistic. This method is widely used in fatigue assessment as well (Madhavan Pillai and Veena, 2006). Considering any one of the above methods to represent hydrodynamic loading, the major task involves computing the stress concentration at critical points (hot spots) for which fatigue damage is to be estimated. In particular, a relationship between the wave height and stress (*S–H* relation) at critical points in the structure needs to be established. Various steps involved in a fatigue assessment procedure are as follows:

- Define wave environment model for wind-generated waves
- Calculate nominal stresses from global structural analysis
- Calculate local stresses at critical points
- Identify appropriate *S–N* curve for each of the detail
- Estimate cumulative fatigue damage or the fatigue strength, using Palmgren–Miner rule
- Alternatively, fatigue strength assessment can be done using fracture mechanics approach
- Perform fatigue safety check using appropriate fatigue design factors (FDF)

A practical procedure should effectively integrate all the above aspects. One of the commonly used methods to study fatigue reliability is *S–N* curve approach, which is discussed in the successive section.

3.16 *S–N* CURVE AND FATIGUE DAMAGE

One of the ways of fatigue assessments is based on the direct calculation of fatigue damage or expected fatigue life (Marshall, 1976; Mustafa, 1985). An indirect fatigue assessment can also be performed by limiting the predicted stress range within the permissible stress range. Fatigue demand is stated in terms of the stress ranges that are produced by the cyclic loads, which are subsequently imposed on the structure. Such loads arise essentially from wave amplitudes, wave and wind directions, etc. As such, loads are also dynamic in nature; the response amplitude of the structure will be dynamically excited under such loads. When considering fatigue-inducing stress ranges, one should also consider the stress concentration and their effects on modifying the active stress range. The *S–N* relation represents the number of cycles to failure at various stress ranges. A two-segment *S–N* curve, well known in various fatigue design applications is given by

$$N = AS^{-m}, \quad N \leq N_Q \tag{3.55}$$

$$N = CS^{-r} N \leq N_Q \tag{3.56}$$

$$N' = N\left(\frac{t}{t_R}\right)^{-mq}, t > t_0; \quad N' = N, t \leq t_0 \tag{3.57}$$

where S denotes the stress range, N the number of cycles to failure at this stress range, for a reference thickness t_R, N' denotes the reduced number of cycles to failure after accounting for shell/plate thickness effect, (A, B) denote fatigue strength coefficients, ($-m$, $-r$) are the fatigue strength exponents, which are determined from fatigue tests. For structures in onshore environment, the relationship ($r = m + 2$) is valid. The above equations represent two-segment straight lines in log scale with a discontinuity at $N = N_Q$, where the stress range value is S_Q, as seen in Figure 3.15.

Traditional *S–N* curves used in design have a horizontal segment in Figure 3.2 for $N \geq N_Q$, which signifies an *endurance limit.* This implies that the member can *endure* for all stress range values $S \leq S_Q$, indicating no fatigue damage to occur in this range. In the early years of offshore structural design, the conservative approach was used by ignoring any endurance limit; the reason stated was to account for corrosion fatigue in the marine environment (Okrent, 1987; Patin Stanislav, 1999). Eventually, with the accumulation of large amount of experimental and field data, the consensus now is to adopt a two-segment *S–N* curve as described above. Tables 3.1 and 3.2 indicate the parameters of *S–N* curve as recommended by ABS code.

As mentioned earlier, *S–N* data are derived from experiments, which do not always include all plate or shell thicknesses of practical interest. *S–N* curves are traditionally

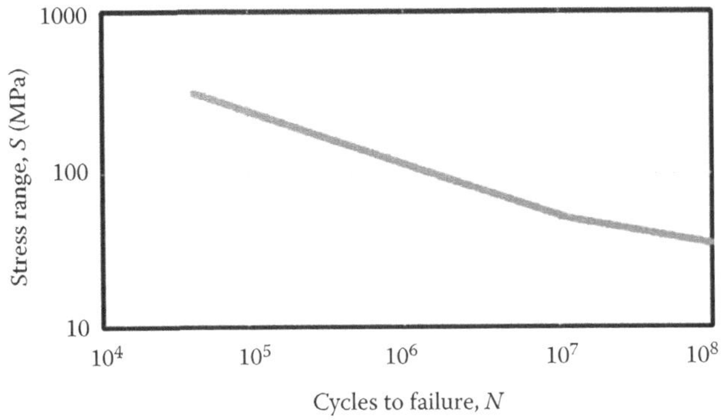

FIGURE 3.15 Two-segment *S–N* curve for different stress ranges.

TABLE 3.1
Parameters for Class "T" Offshore S–N Curves

S–N Curve	A (MPa)	m	C (MPa)	r	N_Q	(MPa)
T(A)	1.46×10^{12}	3.0	4.05×10^{15}	5.0	1.0×10^{7}	52.7
T(CP)	7.30×10^{11}	3.0	4.05×10^{15}	5.0	1.77×10^{6}	74.5
T(FC)	4.87×10^{11}	3.0	–	–	–	–

Note: For service in seawater with free corrosion (FC), there is no change in the curve slope.

TABLE 3.2
Parameters for Plate Thickness Adjustment

(a) Plated Joints

Parameters	European Standard EN 10225: 2001	HSE (1990)	DNV (1992)
q	0.25	0.30	0.0–0.25 (depending on detailed classification)
t_R	22 mm	16 mm	25 mm

(b) Tubular Joints

Parameters	APIRP 2T (1997), APIRP 2A-WSD (2000)	HSE (1990)	DNV (1992)
q	0.25	0.30	0.25 for $SCF < 10.0$ 0.30 for $SCF > 10.0$
t_R	25 mm	16 mm	32 mm

determined by the constant amplitude testing with a large stress ratios (ratio of minimum to maximum stress is taken as 0.5), implying that the stress range considered in the experiments is in the crack-opening mode. A few factors that make the existing *S–N* curve conservative are namely: (i) relaxation of residual stress; (ii) external stress being partly compressive; and (iii) crack-closure effects. These factors would make the actual crack growth lesser than that implied by the stress ranges used in the existing *S–N* curves. While it is good to know that the approach is conservative, the bad news is that they deal with constant amplitude while the stresses are in variable amplitudes due to the nature of environmental loads that offshore structures encounter. Hence, it is recommended that the *S–N* data be used with thickness correction as indicated in Equation 3.57; recommended reference thickness t_R and exponent q are given in Table 3.2 for both plated and tubular joints. An alternate way of thickness adjustment is to define a thickness, which is adjusted for a stress range as given below:

$$s_f = S\left(\frac{t}{t_R}\right)^{-q}, t > t_0; \quad S_f = S, t \le t_0 \tag{3.58}$$

where t is the thickness of the plate.

3.17 ESTIMATE OF CUMULATIVE DAMAGE (LINEAR DAMAGE HYPOTHESIS)

Palmgren–Miner's rule supplements for the constant amplitude, which is not of serious concern when applied to the offshore structures. This actually allows number of different amplitude cycles and the fatigue damage is based on this rule. Fatigue damage of joint, under n cycles of constant amplitude is given by Equation 3.55 and the cycle is simply (n/N). If the joint is under the loading of variable amplitude, then the loading cycles can be divided into groups of approximately equal stress ranges. If there are G such groups with almost equal stress range in a given variable amplitude loading, then let S_g and n_g be the stress range and number of cycles in each group. Fatigue damage of each group is given by

$$D_g = (n_g / N_g) \tag{3.59}$$

$$N_g = A S_g^{-m} \tag{3.60}$$

Miner's rule states that failure under variable amplitude load will occur when the following condition is satisfied:

$$\sum_{g=1}^{G} D_g = 1 \tag{3.61}$$

This is often referred as *fatigue spectrum*. The Palmgren–Miner rule ignores load sequence effects. Further, while applying it to offshore structures, mean stress effects are not taken in to account (Gong et al., 2007). Due to limitations in study of fatigue assessment in offshore structural members under the combined effect of material degradation and load (stress) cycles, few researchers suggested that the Miner rule be used with a damage sum of 0.5 (instead of 1, as seen in Equation 3.61). It is also noted that techniques such as TIG dressing improved the fatigue strength over that of gas-welded specimens.

While determining fatigue assessment of tubular joints, it is necessary to include the following details: (i) standard definition of fatigue failure of tubular joints employed in tests is essential; (ii) size of tubular joints to be examined should be pre-decided as the size has a significant influence on the fatigue strength (for instance, tests show that larger joints have lower fatigue strength); (iii) shape of toe of weld is important for fatigue behavior of welded joints; (iv) post-weld treatment can significantly affect the fatigue strength and should be stated; and (v) influence of a combination of different variables such as seawater corrosion, stress ratio, type of loading, and weld improvement cannot be determined easily and hence should be accounted for indirectly.

3.18 DESIGN *S–N* CURVES

Various classification societies have recommended deign *S–N* curves for both plate-type details and tubular joint details (ABS, AWS, API, Den/HSE, and DNV). Two typical recommendations are given below:

The API-RP2A recommends the following form of S–N curve:

$$N = 2 \times 10^6 \left(\frac{\Delta\sigma}{\Delta\sigma_{ref}} \right)^{-m} \tag{3.62}$$

$$N' = N \left(\frac{t}{t_R} \right)^{-0.25m} \tag{3.63}$$

The Norwegian Standards (NS3472E) recommend the following form:
In air:

$$\text{for } N \leq 10^7, \log N = 12.160.75 \log \frac{t}{32} - 3.0 \log \Delta\sigma \tag{3.64}$$

where t is in millimeters
In water:

$$\text{for } N \leq 10^8, \log N = 12.16 - 0.75 \log \frac{t}{32} - 3.0 \log \Delta\sigma \tag{3.65}$$

It is important to know that the above design S–N curves cannot be applied until suitable stress concentration factors (SCFs) are applied, which are obtained from parametric equations. Such equations have been developed for typical joint configurations using both experimental and finite element modeling; however, they show a lot of discrepancies. A separate section is presented later, wherein SCFs are evaluated experimentally for tubular joints of different configurations.

3.19 FATIGUE ASSESSMENT USING DISCRETE WAVE APPROACH

To enable faster fatigue calculation using discrete wave approach, two versions of the S–H relation, namely a quadratic model and a power form are discussed in the literature (Marshall, 1974; 1976).

3.19.1 S–H Relationship

The most time-consuming part of fatigue assessment is to establish a simple relationship between the stress range (at any point of interest on the structure) and the wave height (i.e., S–H relationship). Quadratic form is given by

$$S = a_1 H + a_2 H^2 \tag{3.66}$$

where the constants a_i can be obtained by a least squares fit to stress values, corresponding to sample wave loadings. It is important to note that a linear contribution is valid only when $a_2 \ll a_1$. It is also possible to use a higher order wave theory, retain the nonlinear terms in the Morison equation, incorporate the variable submergence effect and account for nonlinear soil–pile interaction in steps to estimate the S–H relationship. In view of the above equation, S–N relation can be written as

$$N = Aa_1^{-m}\left[1+\frac{a_2H}{a_1}\right]^{-m} \tag{3.67}$$

As an alternative to the polynomial form in Equation 3.66, one can also use the following form:

$$S = cH^g \tag{3.68}$$

where the constants c and g are obtained by a least squares fit to stress values in the structure corresponding to sample wave loading. Subsequently, S–N relation is written as

$$N = Ac^{-m}H^{-mg} \tag{3.69}$$

According to the Palmgren–Miner rule for cumulative damage, we know that

$$D = \sum \frac{n(S_i)}{N(S_i)} \tag{3.70}$$

where n denotes the number of cycles experienced at stress range S_i, N denotes the cycles to failure corresponding to S_i, and D is the total damage (damage ratio) accumulated due to all the stress cycles experienced by the structure. A member or connection is assumed to have undergone fatigue failure if $D = 1$ as per the Miner hypothesis. The fractional fatigue damage caused due to wave cycles may be written as

$$dD = \frac{dn}{N} \tag{3.71}$$

which implies the detail has undergone dn cycles of a stress range, for which failure occurs at N cycles, according to the S–N data.

3.19.2 Fatigue Damage

Using the above relationships, and assuming the following relationship:

$$\frac{a_2H}{a_1} \ll 1 \tag{3.72}$$

By neglecting higher powers of Equation 3.72 and integrating between the limits $(0,\infty)$, the following S–N relationship is obtained:

$$D = M\frac{(a_1\delta)^m}{A}\left[\Gamma\left(1+\frac{m}{\gamma}\right)+\frac{a_2\delta m}{a_1}\Gamma\left(1+\frac{M}{\gamma}+\frac{1}{\gamma}\right)\right] \tag{3.73}$$

where $\Gamma(p)$ denotes the gamma function defined by

$$\Gamma(P) = \int_0^\infty t^{(p-1)}e^{-t}dt,\ p>0 \tag{3.74}$$

where M is the total number of waves in the sea state under consideration. Similarly, for a two-segment S–N curve, the damage is given by

$$D = M\frac{(a_1\delta)^m}{A}\left[\Gamma\left(1+\frac{m}{\gamma},Z\right)+\frac{a_2\delta m}{a_1}\Gamma\left(1+\frac{m}{\gamma}+\frac{1}{\gamma},Z\right)\right]$$
$$+M\frac{(a_1\delta)^r}{C}\left[\Gamma_0\left(1+\frac{r}{\gamma},Z\right)+\frac{a_2\delta r}{a_1}\Gamma\left(1+\frac{r}{\gamma}+\frac{1}{\gamma},Z\right)\right] \tag{3.75}$$

$$\Gamma(p,Z) = \int_Z^\infty t^{(p-1)}e^{-t}dt = \Gamma(p)-\Gamma_0(p,Z);\quad p>0 \tag{3.76}$$

$$\Gamma_0(p,z) = \int_0^z t^{(p-1)}e^{-t}dt,\quad p>0 \tag{3.77}$$

(*Note*: $\Gamma_0(a, z)\Gamma(a)$ is called the incomplete gamma function.)

$$z = \left(\frac{S_Q}{\delta}\right) \tag{3.78}$$

The long-term fatigue damage due to a sea state modelled using the exponential distribution can be obtained by setting $\gamma = 1$ in Equations 3.73, 3.75 and denoting the δ parameter in terms of the slope of the H–N plot. For the S–H relationship in power form, as given by Equation 3.68, damage expression for a two-segment S–N curve is given by

$$D = M\delta^{gm}\frac{C^m}{A}\Gamma\left(1+\frac{gm}{\gamma},z\right)+M\delta^{gr}\frac{C^r}{A}\Gamma\left(1+\frac{gr}{\gamma},z\right) \tag{3.79}$$

For the single segment S–N curve, damage estimate is given by

$$D = M\delta^{gm}\frac{C^m}{A}\Gamma\left(1+\frac{gm}{\gamma}, z\right) \tag{3.80}$$

3.20 SIMPLIFIED FATIGUE ASSESSMENT METHOD

If we assume the long-term stress range distribution as a two-parameter Weibull distribution given below

$$P(s) = 1-\exp\left[-\left(\frac{S}{\delta}\right)^{\gamma}\right], \quad S>0 \tag{3.81}$$

Long-term damage for a single segment S–N curve is given by

$$D = M\frac{\delta^m}{A}\Gamma\left(1+\frac{m}{\gamma}\right) \tag{3.82}$$

Long-term for two-segment S–N curve is given by

$$\begin{aligned} D &= N_T\frac{\delta^m}{A}\Gamma\left(1+\frac{m}{\gamma}, z\right)+N_T\frac{\delta^r}{C}\Gamma_0\left(1+\frac{r}{\gamma}, z\right) \\ D &= N_T\frac{\delta^m}{A}\Gamma\left(1+\frac{m}{\gamma}, z\right)+N_T\frac{\delta^r}{C}\Gamma_0\left(1+\frac{r}{\gamma}, z\right) \end{aligned} \tag{3.83}$$

The above method gives simple estimate of fatigue damage as neither loading nor stress analysis (to obtain stress range information) is required; instead, a probabilistic model for the stress range is assumed. Fatigue life estimates, calculated from the above equations may vary from that of the earlier with respect to the shape parameter γ. To reduce this ambiguity, one needs to calibrate the result by comparing it with that of detailed analysis for a site-specific case of offshore structure.

3.20.1 Effect of Dynamic Amplification

If the structural frequencies are capable of causing appreciable dynamic amplification to the response, as in the case of large compliant offshore structures, the S–H relationship needs to be modified to account for this effect. It is interesting to note that amplification is significant only in the fundamental mode (Williams and Rinne, 1976). To estimate the amplification factor, peak wave frequency (T_P) is used as the load frequency and a linear relationship between H and log T_P is used, which is derived from the wave scatter diagram. Dynamic response is then written as

$$\sigma = f(H)q(H) \tag{3.84}$$

where $q(H)$ is the typical dynamic amplification factor and $f(H)$ is the quasi-static response.

3.21 SPECTRAL FATIGUE ANALYSIS OF OFFSHORE STRUCTURES

It is well known that the response of a linear system subjected to random loads with Gaussian distribution is also Gaussian. Although the Gaussian sea surface model forms the basis for spectral analysis, the load and response can become non-Gaussian unless a consistent linearization procedure is adopted. Stress response of the offshore structure is often assumed to be a narrowband process for which both the zero-up crossings and peaks follow Rayleigh distribution. If T^* denotes the duration of a storm, which is assumed to be a stationary, cumulative fatigue damage, on the basis of the Miner rule is given by

$$D = \frac{T^*}{T_0}\left[\int_0^{S_Q} \frac{p(S)}{N(S)}\,ds + \int_{S_Q}^{\infty} \frac{p(S)}{N(s)}\,ds\right] \tag{3.85}$$

where $p(S)$ is the Rayleigh distribution for stress range in a narrowband process. Pdf of the stress range, S is given by

$$p(S) = \frac{S}{4\mu_0}\exp\left(\frac{-s^2}{8\mu_0}\right) \tag{3.86}$$

where μ_0 is the zeroth spectral moment of the stress process $s(t)$. Comparing it with Equation 3.85, it is observed that $M = T^*/T_0$, both time measures expressed in seconds. Damage estimate for a two-segment S–N curve for the ith short-term sea state is given by

$$D_i = \frac{T_I^*}{T_{0I}}\left[\frac{1}{A}(\mu_{0i})^{m/2}(8)^{r/2}\Gamma_0\left(1+\frac{r}{2},z\right) + \frac{1}{C}(\mu_{0i})^{r/2}(8)^{r/2}\Gamma_0\left(1+\frac{r}{2},z\right)\right]z$$

$$= \left(\frac{S_Q}{\sqrt{8\mu_{0i}}}\right)^{\gamma} z = \left(\frac{S_Q}{\sqrt{8\mu_{0i}}}\right)^{\gamma} \tag{3.87}$$

The corresponding expression for a one-segment S–N curve is given by

$$D_i = \frac{T_i^*}{T_{0i}}\frac{1}{A}(\mu_{0i})^{m/2}8^{m/2}\Gamma\left(1+\frac{m}{2}\right) \tag{3.88}$$

In order to obtain the total cumulative damage over a given period of time, one may sum up the fractional damages occurred due to various discrete short-term sea states during the period, say T. For this purpose, let p_i denote the fraction of time the ith short-term sea sate prevail over a total duration of T, so that $T_i^* = p_i T$. Since the above estimates of damage are based on a narrowband model, there is a need to apply wide band correction. Let the wide band correction factor for the ith sea state be λ_i. Then, for the two-segment S–N curve the total fatigue damage during the long term is given as

$$D = \sum_{i=1}^{ns} \lambda_i D_i = T \sum_{i=1}^{ns} \left\{ \frac{\lambda_i p_i}{T_{0i}} \left[\frac{1}{A} (\mu_{0i})^{m/2} (8)^{m/2} \Gamma\left(1 + \frac{m}{2}\right) + \frac{1}{C} (\mu_{0i})^{r/2} (8)^{r/2} \Gamma\left(1 + \frac{m}{2}, z\right) \right] \right\} \quad (3.89)$$

3.22 SHORT-TERM FATIGUE DAMAGE

The power spectral density of wave surface elevation for low-to-moderate sea states cannot strictly be assumed to be narrowband, whereas such an assumption would work better for extreme sea states. The stress response due to broadband sea state will continue to be broadband particularly when the dynamic response is not dominated by large dynamic amplifications in any given natural modes of vibration. Thus, wave-induced stresses in offshore structural members that are exposed to low-to-moderate sea states exhibit wide/broadband, which cannot be ignored in fatigue calculations. They are likely to have significant humps near the wave spectral peak as well as in the vicinity of the fundamental frequency of the structure. For such a case, closed-form expressions for stress cycles and hence the fatigue damage, derived from spectral analysis will not be very accurate (Wirsching and Light, 1980). A general approach to the wide band correction is to assume a suitable counting algorithm to estimate the stress cycles in the time history of stresses generated by a time domain analysis. Subsequently, this is used to estimate the stress range history to obtain the damage. Rain flow counting method is one of the methods popular in offshore fatigue calculations. Given a stress spectrum, which is wide band, a sample stress history can be generated using computer simulation. The rain flow method can then be used to count the stress cycles in the simulated record. Following such an approach, narrowband and wide band are connected by the following relationship:

$$D_{WB} = \lambda D_{NB} \quad (3.90)$$

where D_{WB} denotes the damage in a wide band process obtained from time domain analysis, D_{NB} denotes the damage obtained by making narrowband assumption, and λ is a correction factor. D_{WB} can be obtained from a rain flow analysis while D_{NB} has a closed form expression, as discussed above. In order to establish suitable values for λ, a few typical spectral shapes have been considered and the variation of λ with the spectral width parameter ε and the exponents in the S–N curve are studied by various researchers. The Wirsching rain flow factor λ is given by

$$\lambda(m, \varepsilon) = a + (1 - a)(1 - \varepsilon)^b \quad (3.91)$$

$$a = 0.926 - 0.033m$$

$$b = 1.587m - 2.3.2$$

Note that m in the above expression refers to the exponent of the S–N curve in the low cycle range.

There are many sources of uncertainty in fatigue analysis, such as wave loads, linearization of wave loads, inaccuracies in SCFs, and fatigue design curves. Fatigue damage estimates are sensitive to the errors introduced due to these factors. Thus, the safety check expression should be used, which is given by

$$D \leq \frac{1.0}{FDF} \tag{3.92}$$

where FDF (>1) denotes the fatigue design factor; it is implied that the total damage D is cumulative over the design life T of the structure. Alternatively, if the calculated long-term fatigue damage, say D_0, corresponds to a year, then the calculated fatigue life T_f is given by

$$T_f = \frac{1.0}{D_0} \tag{3.93}$$

Then, the member or the joint is considered to be safe against fatigue failure if the following condition is satisfied:

$$T_f \geq T_X FDF \tag{3.94}$$

Appropriate values of FDF are available in international codes as design guidelines.

3.22.1 Evaluation of Damage Integrals

The integrand in the fatigue damage integral contains basically N, the number of cycles to failure, and variants of exponential distribution for the probability of either density/distribution function. Consider an integral of the following form:

$$D = \int dD = \int \frac{dn}{N} \tag{3.95}$$

where dn may be written as

$$dn = Mfds \tag{3.96}$$

The variable M in the above equation denotes the number of waves in a long-term or short-term wave climate. Accordingly, $f(S)$ would denote either the probability distribution or pdf of the variable S, depending on the approach used to quantify it. As an example consider the form:

$$f(s)=\left(\frac{1}{\delta}\right)\left(\frac{-s}{\delta}\right)^{\gamma-1}\exp\left\{-\left(\frac{s}{\delta}\right)^{\gamma}\right\} \tag{3.97}$$

This can be obtained from the Weibull distribution function. Damage expressions in Equation 3.95 will consist of integrals of the following form:

$$I=\int_{0}^{S_q}\frac{(1/\delta)s/\delta^{\gamma-1}\exp\{-S/\delta\}^{\gamma}}{AS^{-m}}+\int_{S_q}^{\infty}\frac{(1/\delta)(S/\delta)^{\gamma-1}\exp(-S/\delta)^{\gamma}}{AS^{-m}}ds \tag{3.98}$$

Let $t=(S/\delta)^{\gamma}$. Then, using gamma functions, the above integrals may written as

$$\Gamma(p,Z)=\int_{z}^{\infty}t^{(p-1)}e^{-1}dt=\Gamma(p)-\Gamma_0(p,Z);\quad p>0 \tag{3.99}$$

$$\Gamma_0(p,Z)=\int_{0}^{z}t^{(p-1)}e^{-t}dt,\quad P>0 \tag{3.100}$$

$$Z=\left(\frac{s_Q}{\delta}\right)^{\gamma} \tag{3.101}$$

where $\Gamma_0(p, Z)\Gamma(a)$ is called the incomplete gamma function. Both $\Gamma_{0(a,}$ $_{z)}$ and $\Gamma(a)$ can be evaluated using MATLAB.

3.23 UNCERTAINTIES IN FATIGUE RELIABILITY

It is known that the various inputs in fatigue damage estimates, as presented above, are prone to uncertainties/errors due to the following reasons:

i. Basic fatigue data, which is used in developing the S–N curves is a source of large scatter.
ii. Fatigue damage model, especially under random stresses is approximate.
iii. Defects in discontinuities at the welded joints complicate the process of crack initiation and propagation. This is also an aspect that relates to the fatigue damage model.
iv. Statistical models used in ocean wave climate modeling have inherent errors.
v. The force/load model for estimating the wave and current forces are only approximate and often empirical.

vi. The hot-spot stresses used in fatigue damage calculations contain uncertainties, because a variety of stress analysis procedures, each having different accuracy, are adopted in practice. This relates to errors in structural modeling for obtaining nominal stresses, the evaluation of hot-spot stresses using SCFs, etc.
vii. Influence of corrosion and cathodic protection on fatigue strength is not well understood yet.

3.24 LOGNORMAL FORMAT FOR FATIGUE RELIABILITY

According to the Palmgren–Miner rule, fatigue failure occurs when $D = 1$. The stress range S may be assumed to relate to the actual stress S_a as

$$S_a = BS \tag{3.102}$$

where B accounts for modeling errors in obtaining the stresses. Assuming B with lognormal distribution and S being a random variable, damage estimate can be expressed as (Wirsching 1984)

$$\Delta = \frac{n}{A} E(S^m) \tag{3.103}$$

where the random variable Δ replaces D to denote fatigue damage, and $E(.)$ denotes the expectation operator. The damage expression in the above equation can be rewritten as

$$\Delta = \frac{TB^m\Omega}{A} \tag{3.104}$$

where Ω is defined as the stress parameter and is given by

$$\Omega = f_0 E(S^m) \tag{3.105}$$

In the above equation, the average frequency of stress cycles is defined as

$$f_0 = \frac{n_T}{T} \tag{3.106}$$

where T denotes the time duration over which n_T stress cycles have occurred. The stress parameter Ω in Equation 3.105 can be derived for a chosen probabilistic model. Denoting the time to fatigue failure by T, using the damage expression, we get

$$T = \frac{\Delta A}{B^m\Omega} \tag{3.107}$$

Probability of fatigue failure of a joint may be written as

$$p_f = P(T \leq T_s) \tag{3.108}$$

where T_s denotes the intended service life of the structure. For a log-normal format probability of fatigue failure p_f is given by

$$p_f = \Phi(-\beta) \tag{3.109}$$

where $\Phi(.)$ denotes the standard normal distribution function (see Appendix), and β is defined as the safety index, which is given by

$$\beta = \frac{ln(\tilde{T}/T_s)}{\sigma_{lnT}} \tag{3.110}$$

where $\tilde{T}$ denotes the median value of T and is equal to

$$\tilde{T} = \frac{\tilde{\Delta}\tilde{A}}{\tilde{B}^m \Omega} \tag{3.111}$$

Note that the Equations 3.111 and 3.107 are similar with the random variables replaced by their median values. Further,

$$\sigma_{lnT} = ln\left[(1+c_{\Delta}^2)(1+c_A^2)(1+c_B^2)^{m^2}\right]^{1/2} \tag{3.112}$$

where Cs denote the COV of the respective random variables.

3.25 TUBULAR JOINTS: EXPERIMENTAL AND ANALYTICAL INVESTIGATIONS

Tubular joints are of utmost importance in offshore structures, with particular respect to fatigue failure. As discussed earlier, determining SCF involves a lot of uncertainties, which lead to difficulties in reliability analysis (or in estimating probability of failure, to be very specific). In the past, it is observed that studies have mostly focused on the estimation of ultimate strength or SCFs rather than the detailed study of joint behavior. Detailed investigations of T joint under axial compressive loads and out-of-plane bending and K joint under axial loading with SCF is presented; the failure pattern, obtained by experimental and numerical investigations is discussed. Failure pattern, ultimate strength, and the stress concentration factors of unstiffened T joints and K joints under axial brace compression loading is discussed. Rohit (2015) reported the numerical modeling overall joint using the Sesame Genies software package. Nominal diameter and thickness of the chord of the unstiffened

T joints are 168.3 and 7.11 mm, respectively. Corresponding values of the brace are 114.3 and 6.02 mm, respectively. Numerical analysis has simulated the behavior of the joint as observed in the experiment. Based on the preliminary experimental investigations, it is seen that the unstiffened joints are stiffer in ovalization mode in comparison to flexural deflection. The stress concentration factor (SCF) is also predicted by the numerical analysis, which is in close agreement with that measured from the experimental investigations; a comparison is also made with that of the established parametric equations. Figure 3.16 shows the nomenclature of the joint, used in the study.

Following nomenclature is important: θ is brace included angle; g is the gap between braces; t is the brace wall thickness at intersection; T is the chord wall thickness at intersection; d is the brace outside diameter; D is the chord outside diameter; and L is the chord length. The following relationships are useful.

$$\tau = \frac{t}{T}, \beta = \frac{d}{D}, \gamma = \frac{D}{2\mathrm{T}}, \alpha = \frac{2\mathrm{L}}{D}, \zeta = \frac{g}{D} \quad (3.113)$$

Tubular joints are connections welded at the interface between tubular members, consisting of a weld deposit, heat-affected zone, and immediately adjacent base metal. The main member is denoted as a chord and the secondary as a brace or branch (refer to Figure 3.16). A joint without any reinforcement is called an unstiffened joint and that provided with internal rings welded to the inside of the chord at

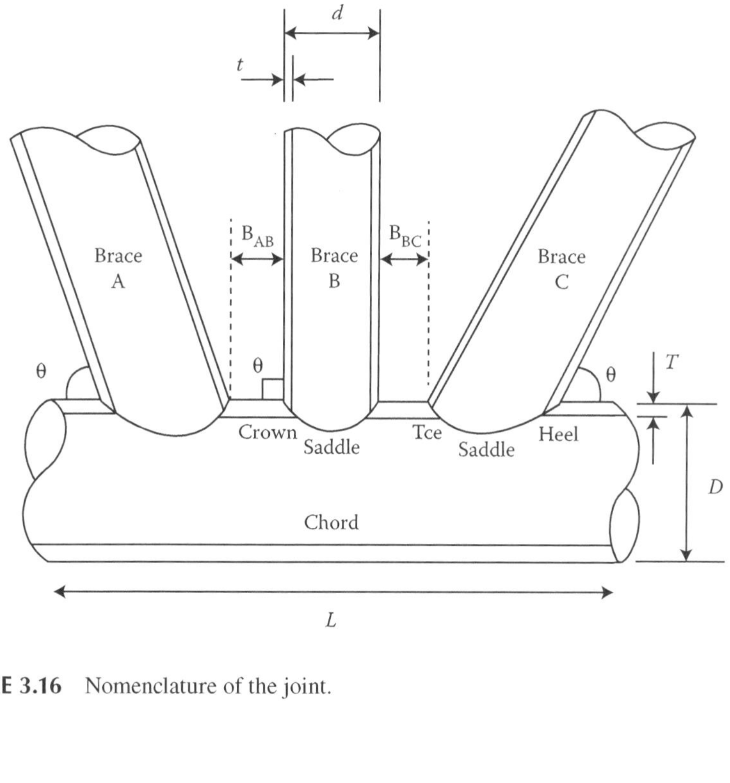

FIGURE 3.16 Nomenclature of the joint.

the intersection is called an internally ring-stiffened joint. Braces produce a high membrane and shell bending stresses in the chord shell, which results in nonuniform stress distribution at the intersection; this amounting to a high stress concentration at the connection and also causing change in geometry at the welds. Conventionally, for tubular joints, SCF is defined as the ratio of maximum stress at intersection to nominal stress on the brace. Because of the cyclic nature of wave loading, tubular joints are most vulnerable to fatigue damage and hence fatigue performance assumes significance in the design of tubular joints. As the fatigue performance of tubular joints depends on SCF, accurate computation of stress concentration is important not only in the design of tubular joints but also in reliability estimates.

Thanks to the extensive studies reported in the literature on unstiffened tubular simple planar joints of T, K, and Y shapes, parametric equations to determine SCF, ultimate strength and fatigue life are available through international codes. Furthermore, recent studies are reported on behavior of tubular joints under combined loads to study the crack propagation at welded intersections. As observed, these studies are mainly focused on the estimation of the SCF, investigation on fatigue behavior, and determination of the ultimate strength of the unstiffened joints of various geometric configuration and planes. But an insignificant attempt is made to study the behavior of unstiffened tubular joints under axial brace compression, loaded up to ultimate failure load. This behavior is necessary to understand the reasons for inaccurate fatigue failure using the conventional methods like SCF, etc. To be more specific, parameters that contribute to the failure, failure pattern under axial brace compression till ultimate load, etc. are important to understand the contributors of uncertainty in reliability estimates of tubular joints. Figure 3.17 shows various geometric configurations of tubular joints, considered for experimental and analytical investigations (Rohit, 2015).

3.25.1 Fatigue Life Estimate of Tubular Joints

Steel members subjected to sufficiently large fluctuating tensile stresses will develop small crack-like defects, which may eventually lead to failure of the member; the crack becomes sufficiently large to cause fracture at the defective location. For structures subject to constant amplitude stress fluctuations, problems of this nature may be avoided by keeping the cyclic stress below an endurance limit, preventing the

FIGURE 3.17 T joint and K joint fabricated for the study.

inception of the initial defect. Offshore structures, however, are exposed to a mixture of large and small stress ranges making this approach impractical. Design of tubular joints against failure is based on *S–N* curve approach (API; AWS 1972). Other methods for determining the fatigue life of a tubular joint is based on fracture mechanics. Hot-spot stress is the region where fatigue cracking is most likely to initiate due to a stress concentration. For most simple joint geometries and loadings, hot spot will be located at either the saddle or crown. However, studies of the measured stress around the brace/chord periphery indicate that hot-spot stress may be located at an interim position for some geometry. A tubular joint fails when one of the conditions is satisfied namely: (i) stress in material reaches elastic limit; (ii) stress in the material reaches yield strength; (iii) first crack is detected in the tension joint; (iv) maximum load capacity of the joint is reached in compression, which can result in large deformation (Srinivasan and Subarata, 2012). In specific terms, T joints under axial loads fail by ovalization and punching shear (Graff, 1981). For tubular connection with $\beta < 0.3$, failure occurs by punching shear, which results in punching in or pulling out the plug from the side of the chord. When $\beta > 0.8$, the chord fails by collapse. For the range in between, one should estimate the interaction of punching shear and general chord collapse to decide the failure pattern. It is also important to note tubular joints fail in multiple modes, which impose challenges in estimating the probability of failure namely: (i) local failure of chord; (ii) general collapse of the chord; (iii) unzipping or progressive failure of the weld; (iv) fracture and delaminating, and (v) fatigue. Based on the finite element analysis using shell elements, parametric equations are developed to estimate remaining life of the joint (more toward reliability assessment) rather than designing a tubular joint (Rajashekar and Ellingwood, 1993; Mihailidou et al., 2012).

3.26 BEHAVIOR OF T JOINTS UNDER AXIAL LOADS

Specimens of *T* joints are fabricated to investigate them under different loading conditions namely: (i) axial loading; (ii) out-of-plane bending and K joints are investigated under axial load only. Geometric parameters for the test specimens are chosen to be within the acceptable range namely: β from [0.2–1.0]; τ from [0.2–1.0]; γ from [8–32]; α from [4–40]; θ from [20–90°]; and ζ (gap) from [$-0.6\beta/\sin\theta$ to 1.0] (API RP2- WSD, 2001). Figure 3.18 show the geometric configuration of the joints investigated.

Welding of the joints is carried out as per API recommendations and also tested by magnetic particle test for any defects. For welded joints, improvement factors on fatigue performance can be obtained by a number of methods. For example, controlled burr grinding of the weld toe, hammer peening, or as-welded profile, etc. to produce a smooth concave profile, which blends smoothly with the parent metal. But for current experimental investigations, no special grinding is done. Experimental setup with the joint in place under the loading frame is shown in Figure 3.19.

An existing 200-ton resisting wall is used as a reaction wall for the applied loads. Stiffened I section is placed between the loading jack and the reaction wall to have a smooth loading surface. *A-frame* is used for fixing the edges of the chord member. The A-frame is locked by the torque machine to the ground to avoid any moment

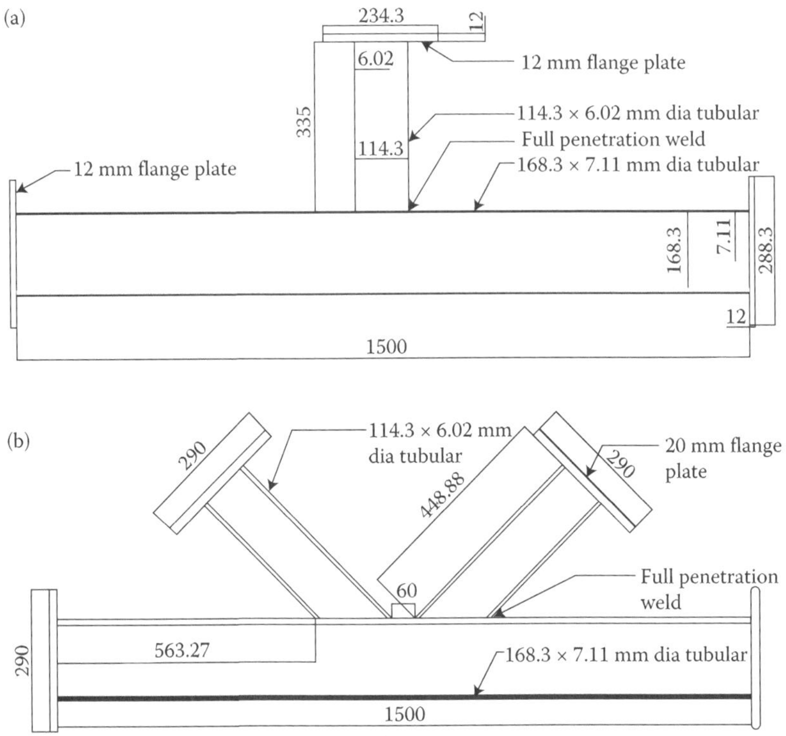

FIGURE 3.18 Geometric details of (a) T joint and (b) K joint considered for the study.

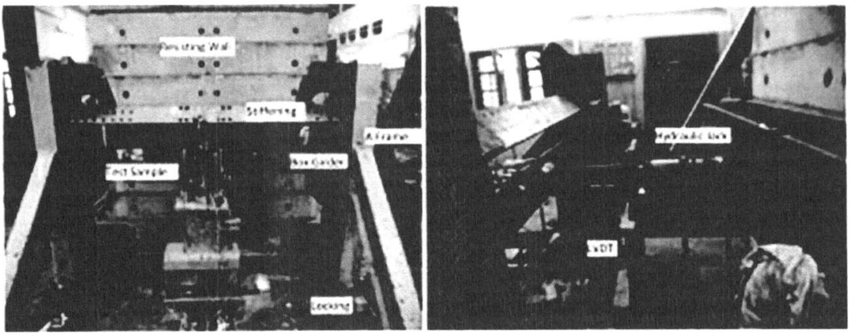

FIGURE 3.19 Experimental setup with components.

at base of the frame. To determine the hot-spot stresses, rosette strain gages (0–45–90°) are used to measure deflection; linear variable differential transformer is used to measure deformation. Catman Data Acquisition software is used for data acquisitions. Figure 3.20 shows the sample fixed with the strain rosette.

Strains are measured on physical models using strain gauge rosettes or a single strain gauge, which is placed perpendicular and parallel to the joint intersection

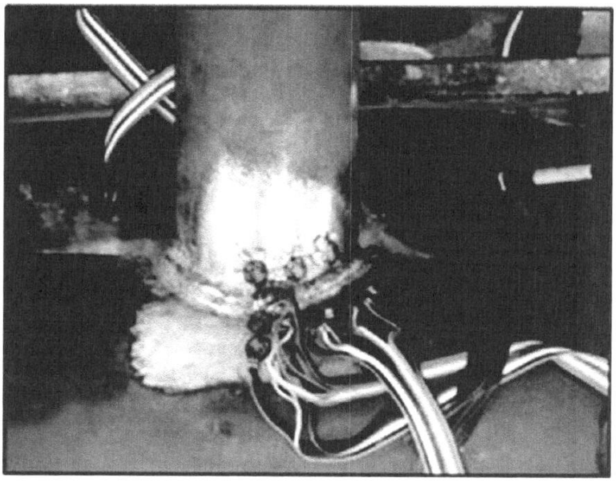

FIGURE 3.20 Sample fixed with strain rosette.

or weld toe. From the measured strain values, maximum principal stresses are given by

$$\sigma_1 = E\left[\frac{(\varepsilon_a + \varepsilon_c)}{2(1-\nu)} + \frac{\sqrt{(\varepsilon_a - \varepsilon_c)^2 + (2\varepsilon_b - \varepsilon_a - \varepsilon_c)^2}}{2(1+\nu)}\right] \tag{3.114}$$

$$\sigma_2 = E\left[\frac{(\varepsilon_a + \varepsilon_c)}{2(1-\nu)} - \frac{\sqrt{(\varepsilon_a - \varepsilon_c)^2 + (2\varepsilon_b - \varepsilon_a - \varepsilon_c)^2}}{2(1+\nu)}\right] \tag{3.115}$$

where σ_1, σ_2 are maximum and minimum principal stresses; ε_a, ε_b, ε_c are strain values measured in at 0°, 45°, 90°, respectively; E is the modulus of elasticity of the material and ν is the Poisson's ratio. During the conduct of the experiments, extra stiffening is provided between the frames to avoid any unwanted damage to the frames and samples; this is also required to maintain the fixity condition at both the ends. Figure 3.21 shows the test set up for the T joint and Table 3.3 shows the details of T joints under investigation.

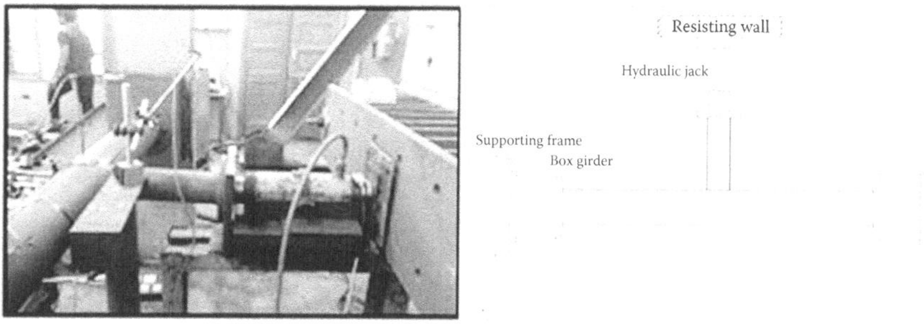

FIGURE 3.21 T joint under axial load set up.

TABLE 3.3
Details of Parameters of T Joint under Axial Loading

S. No	Specimen No.	Chord (mm)	Brace (mm)	β	τ	γ	A
1	T-1	168.3 × 7.11	114.3 × 6.02	0.68	0.84	11.84	4.46
2	T-2	168.3 × 7.11	114.3 × 6.02				
3	T-3	168.3 × 7.11	114.3 × 6.02				

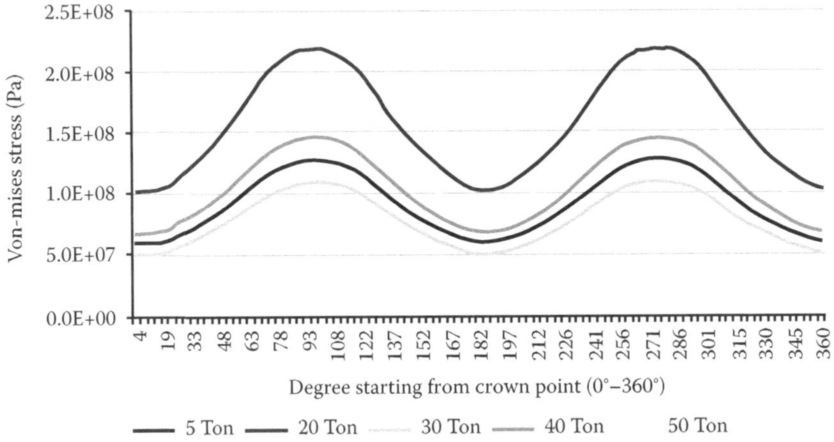

FIGURE 3.22 HSS variation across the profile for tubular joint under axial loading.

Numerical analysis was also done for the T joint under axial load to compare the behavior obtained from that of the experimental investigations. SCF are obtained from the numerical analysis. Figure 3.22 shows its variation across the profile of the tubular joint under axial load.

API RP 2A recommendations are generally used to calculate SCF for offshore tubular joints. Efthymiou parametric SCF equations are given in API RP 2A,

TABLE 3.4
SCF Equation and Parameter Validation for T Joint under Axial Loading

Location	Equation	Short Chord Correction
Chord saddle	$\gamma\tau^{1.1}(1.11 - 3(\beta - 0.52)^2)(\sin\theta)^{1.6}$	F1
Chord crown	$\gamma^{0.2}\tau(2.65 - 5(\beta - 0.65)^2) + \tau\beta(0.25\alpha - 3)\sin\theta$	None
Brace saddle	$1.3 + \gamma\tau^{0.52}\alpha^{0.1}(0.187 - 1.25\beta^{1.1}(\beta - 0.96))(\sin\theta)^{2.7-0.01\alpha}$	F1
Brace crown	$3 + \gamma^{1.2}(0.12\exp(-4\beta) + 0.011\beta^2 - 0.045) + \beta\tau(0.1\alpha - 1.2)$	None

which would be used in this study to calculate the SCF for tubular joints T/Y under axial load with chord ends fixed. Table 3.4 shows the SCF equation and parameter validation.

Parameter	Value	Valid Range	In Valid Range?
α	17.83	$4 \le \alpha \le 40$	Yes
β	0.68	$0.2 \le \beta \le 1.0$	Yes
γ	11.84	$8 \le \gamma \le 32$	Yes
τ	0.85	$0.2 \le \tau \le 1.0$	Yes
θ	90.00	$20° \le \theta \le 90°$	Yes

To find the SCF experimentally, strain values measured are converted to appropriate stress values to compute the applied stress, which is given by

$$Applied\ stress\ (nominal\ stress) = \frac{Applied\ load\ on\ brace}{Area\ of\ brace} \tag{3.116}$$

$$SCF = \frac{Hot\ spot\ stress}{Nominal\ stress} \tag{3.117}$$

One of the T joint is tested under yield to determine the stress behavior at the Chord saddle point, which is the hot-spot stress point. Figure 3.23 shows the variation of SCF with that of the applied axial load. Figure 3.24 compares the SCF obtained from the experiments for the chord and brace. Figure 3.25 compares the SCF variation obtained from the numerical analyses. Figures 3.26 and 3.27 show the comparison of SCF obtained from experimental, numerical, and those obtained

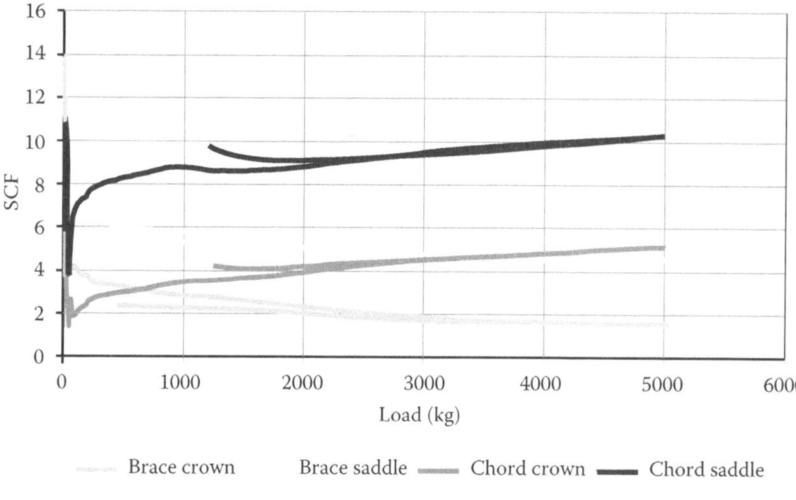

FIGURE 3.23 SCF variation of T joint under axial loading.

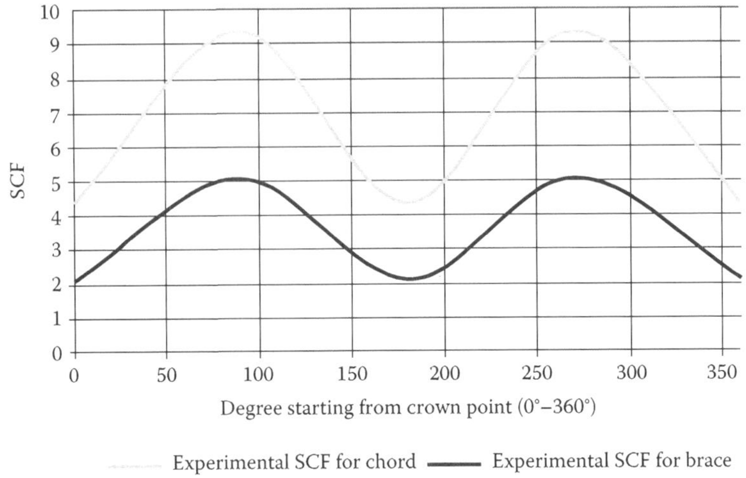

FIGURE 3.24 SCF variation around the T joint under axial load: experimental observations.

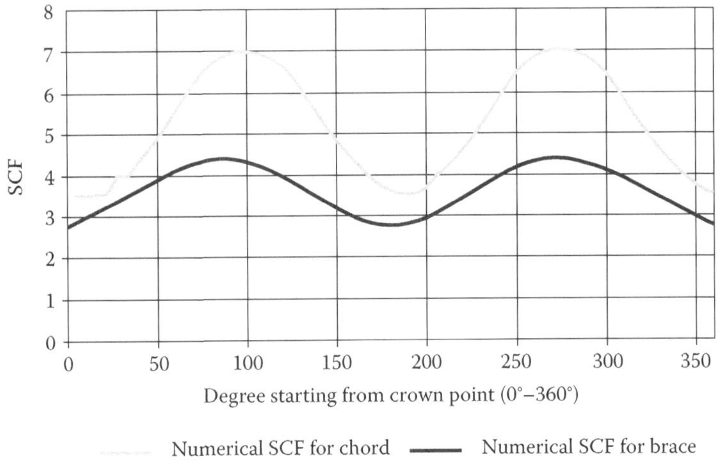

FIGURE 3.25 SCF variation around T joint under axial load: numerical analyses.

using the parametric equations for T joint, chord member, and brace member under axial load, respectively.

In order to understand the variations on SCF values, which influence the fatigue life estimates, fatigue assessment of T joint is done using *S–N* curve approach. Design *S–N* curve for tubular joints in air with a chord wall thickness of 16 mm is given by

$$\mathrm{Log}_{10}(N) = \mathrm{Log}_{10}(k1) - m\,\mathrm{Log}_{10}(S) \tag{3.118}$$

where N is the predicted number of cycles to failure under stress range S, k_1 a constant, and m is the inverse slope of the *S–N* curve. Table 3.5 gives the value of the

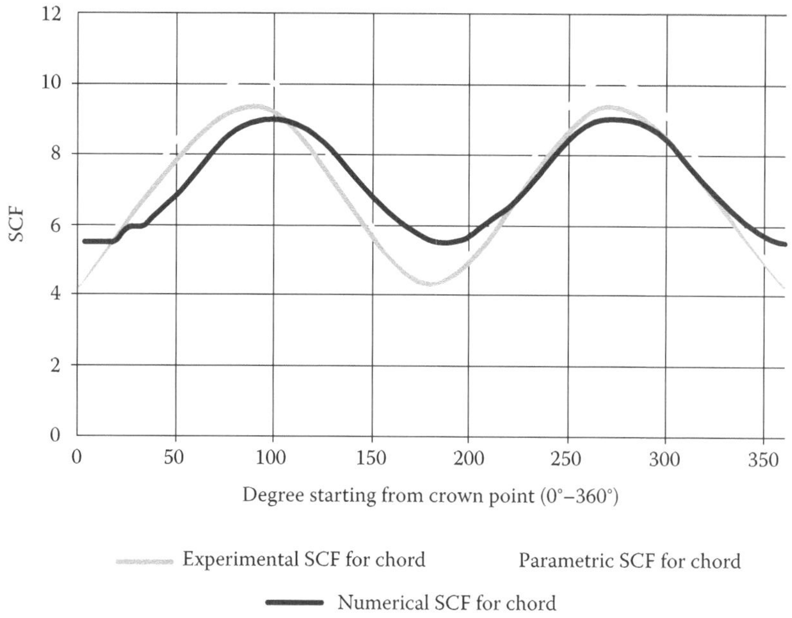

FIGURE 3.26 Comparison of SCF at T joint chord member.

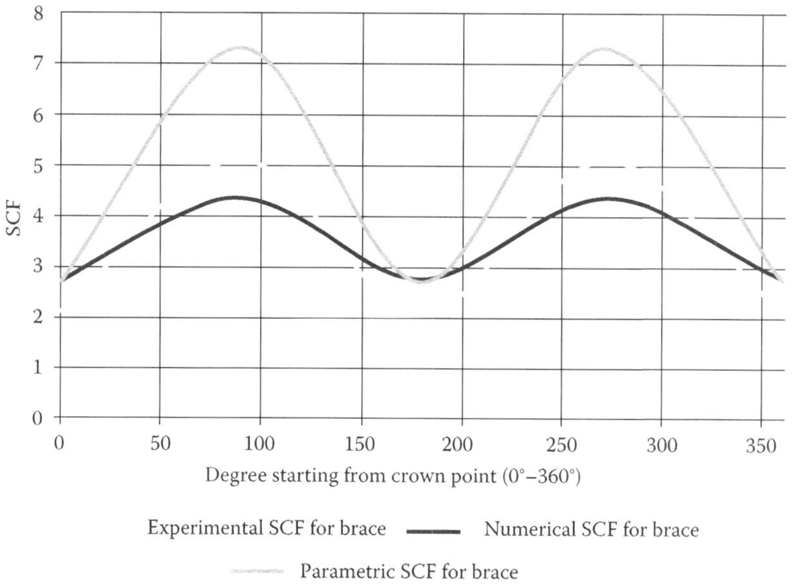

FIGURE 3.27 Comparison of SCF at T joint brace member.

TABLE 3.5
Constant k_1 Value

Curve	$\text{Log}_{10}(k_1)$ When *S* in MPa	*m*
Welded joints (WJ)	12.48	3 for $N < 10^7$
Welded joints (WJ)	16.13	5 for $N > 10^7$

TABLE 3.6
Allowable Number of Cycles for T Joint for Different SCF Values

T Joint			$б_{nom}$	SCF	Δбo	Thickness Correction Δбo	*N* Cycle
Axial	Chord	Experimental	5	9.37	46.85	57.38,154	15,837,383
	saddle	Parametric	5	10.19	50.95	62.40,319	12,313,447
	Brace	Experimental	5	5.1	25.5	31.23,221	98,218,132
	saddle	Parametric	5	7.33	36.65	44.88,865	33,081,887

constant as per the code (API RP2- WSD, 2001). Table 3.6 compares the estimate of number of cycles for T joint, obtained from experimental investigations and parametric methods.

From the above comparison, it can be observed that the fatigue life of a joint significantly increases with the experimental SCF, in comparison to that of parametric equations. Uncertainties with respect to life cycle prediction of offshore structures can be easily seen through this comparison. Table 3.7 shows the fatigue life of a T joint under axial loading.

Considering 500,000 cycles per year, increase in the life of the joint is from 24.62 years to 31.68 years. For all the three joints, the failure observed is a punching failure with the weld intact. Ovalization of the chord is observed till the chord member yielded after which plastic deformation occurred near the weld area resulting in the sudden punching of the joint. Figures 3.28 and 3.29 show the failure of the joint in both the punching shear mode and bulging of joint, indicating a plastic hinge formation. It can be seen from the figures that the influence of the axial load on the chord is extended beyond the brace diameter region, which is important information to account for uncertainties in reliability parameters; Figure 3.30 shows the region of

TABLE 3.7
Fatigue Life of T Joint under Axial Loading

Loading	Method	*N*	N/Year	Fatigue Life (Year)
Axial	Experimental	15,837,383	500,000	31.674,766
	Parametric	12,313,447	500,000	24.626,894
	Experimental	98,218,132	500,000	196.436,264
	Parametric	33,081,887	500,000	66.163,774

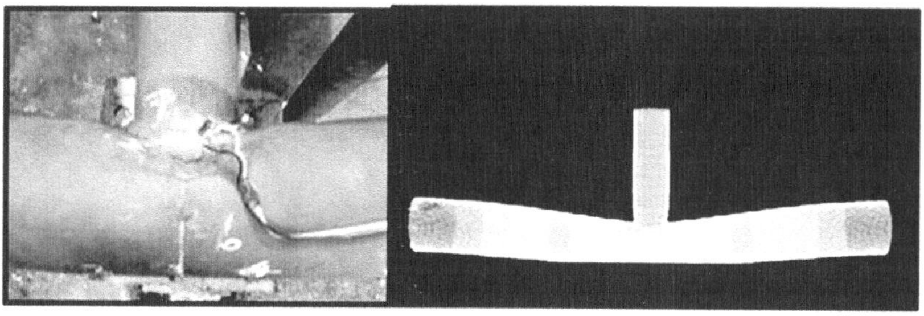

FIGURE 3.28 Failed T joint in punching shear.

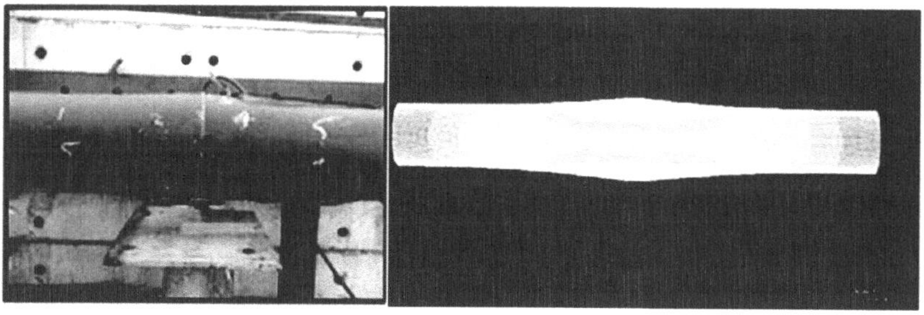

FIGURE 3.29 Failed T joint showing bulging of joint.

FIGURE 3.30 Region of influence of T joint under axial load failure.

influence. Maximum hot-spot stress is observed at the saddle point and minimum at the crown point of the weld.

Table 3.8 shows the comparison of the SCF around the T joint weld under axial loading. It can be seen from the table that the SCF computed from the experimental investigations is less than that of the parametric equations, indicating the necessity of reliability estimates through experimental investigations. The fatigue life of the T

TABLE 3.8
SCF Variation around T Joint Weld by a Different Method

Starting from Crown Point (0°–360°)	Experimental SCF		Parametric SCF		Numerical SCF	
	Chord	Brace	Chord	Brace	Chord	Brace
0	4.35	2.1	4.52	2.71	5.5	2.75
90	9.37	5.1	10.19	7.33	9.05	4.38
180	4.35	2.1	4.52	2.71	5.5	2.75
270	9.37	5.1	10.19	7.33	9.05	4.38
360	4.35	2.1	4.52	2.71	5.5	2.75

joint increases significantly when the experimental SCF values are used; increase in life of the joint is from 24.62 to 31.68 years.

3.27 T JOINT UNDER OUT-OF-PLANE BENDING

Out-of-plane bending is a common situation of members in offshore structures subjected to environmental loads. Under the combination of loads, out-of-plane bending can actually reduce the ultimate strength of the joint. In particular, the SCF and reliability estimates can be influenced significantly. T joints are experimentally investigated to understand the failure behavior and fatigue life estimate of under out-of-plane bending criteria. Table 3.9 shows the details of the T joints considered for investigations.

Figure 3.31 shows the experimental setup for T joint under the loading case of out-of-plane bending.

Figure 3.32 shows the hot-spot stress variations, numerically, while Figure 3.33 shows the variation around the profile of tubular joint.

SCFs for the tubular joints (T/Y) with chord ends fixed and subjected to out-of-plane bending are given in Table 3.10 (APIRP 2A). SCF are also computed from the strain values measured experimentally. Figure 3.34 shows the variations of SCF values of T joint under out-of-plane bending; values are computed from the measured strain.

The average experimental SCF value for a tubular T joint under axial loading condition is 6.29. Figures 3.35 and 3.36 show the hot-spot stresses as stress contour and the variation of SCF around the weld of a T joint, respectively. SCFs for T joints

TABLE 3.9
T Joint under Out-of-Plane Bending

S. No	Specimen	Chord (mm)	Brace (mm)	β	τ	γ	α
1	T-4	168.3 × 7.11	114.3 × 6.02	0.68	0.84	11.84	4.46
2	T-5	168.3 × 7.11	114.3 × 6.02				
3	T-6	168.3 × 7.11	114.3 × 6.02				

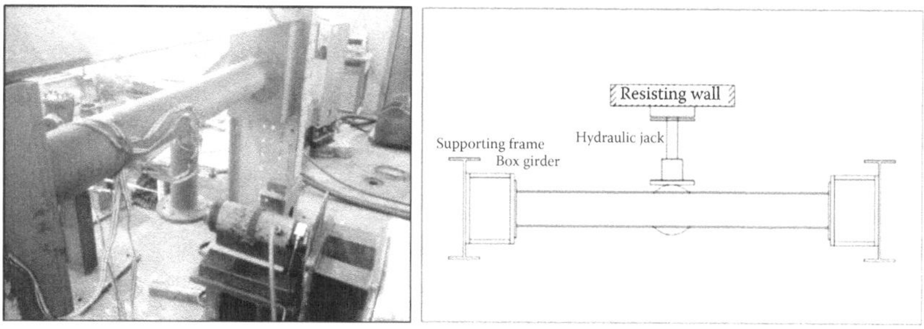

FIGURE 3.31 T joint under out-of-plane bending.

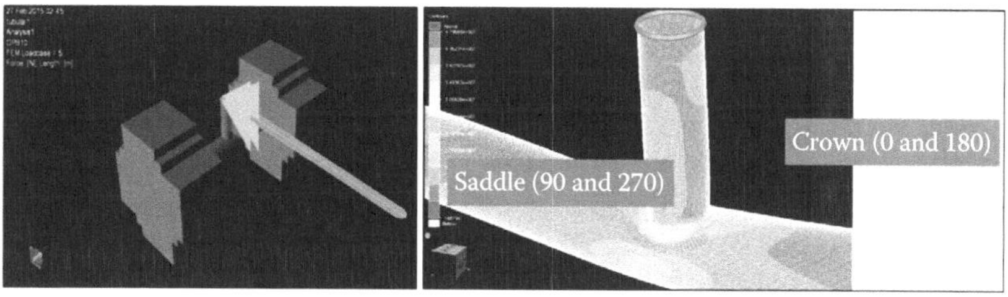

FIGURE 3.32 Hot-spot stress of T joint under out-of-plane bending.

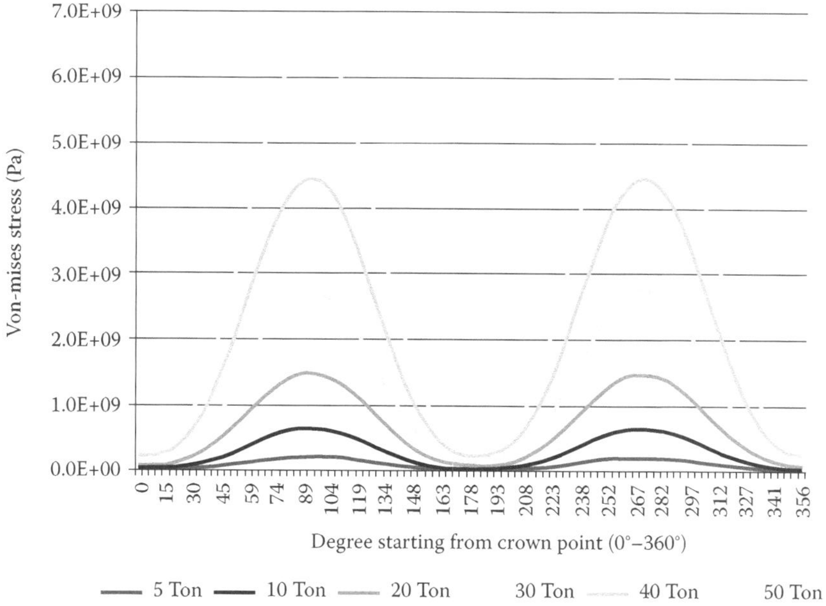

FIGURE 3.33 Hot-spot stress variation around the T joint under out-of-plane bending.

TABLE 3.10
SCF Equation and Parameter Validation for T Joint under OPB Loading

Location	Equation	Short Chord Correction
Chord saddle	$\gamma\tau\beta(1.7 - 1.05\beta^3)(\sin(\theta))^{1.6}$	F3
Brace saddle	$\tau^{-0.54}\gamma^{-0.05}(0.99 - 0.47\beta + 0.08\beta^4)(\gamma\tau\beta(1.7 - 1.05\beta^3)(\sin(\theta))^{1.6})$	F3

Note: Short chord correction factor ($\alpha < 12$).
$F3 = 1 - 0.55\beta^{1.8}\gamma^{0.16}\exp(-0.49\gamma^{-0.89}\alpha^{1.8})$

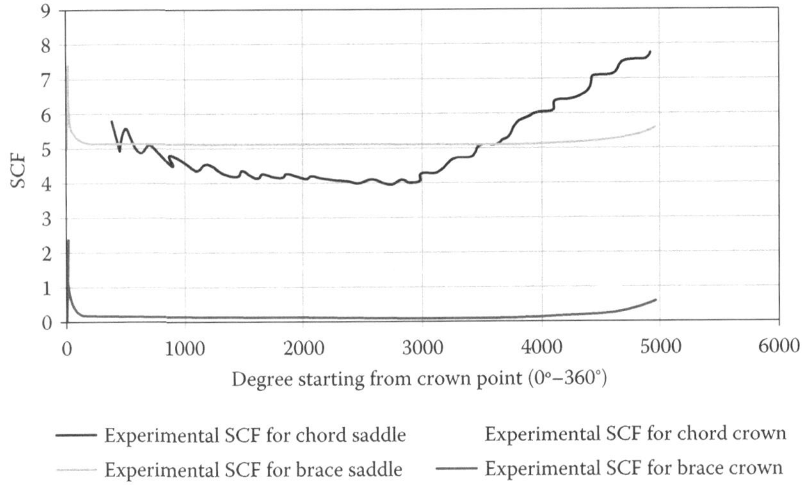

FIGURE 3.34 SCF of T joint under out-of-plane bending (experimental).

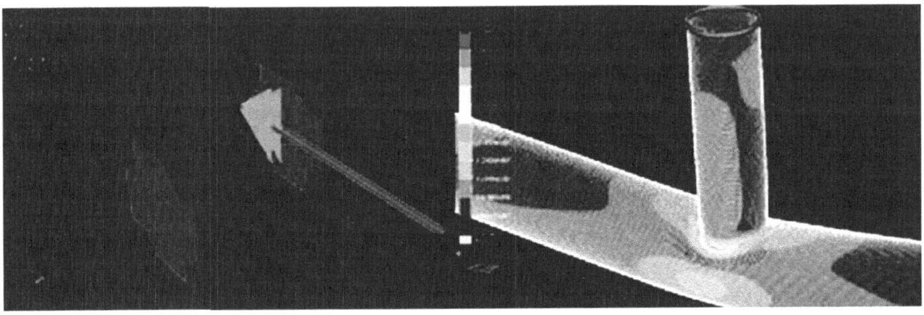

FIGURE 3.35 Hot-spot stress contour in tubular T joint under out-of-plane bending.

under out-of-plane bending are also computed from parametric equations as shown in Table 3.11. Comparison is plotted for the chord member and the brace in Figures 3.37 and 3.38, respectively.

By comparing the above figures, it is seen that the parametric SCF value is higher than that of the experimental and numerical SCF values; both numeric and

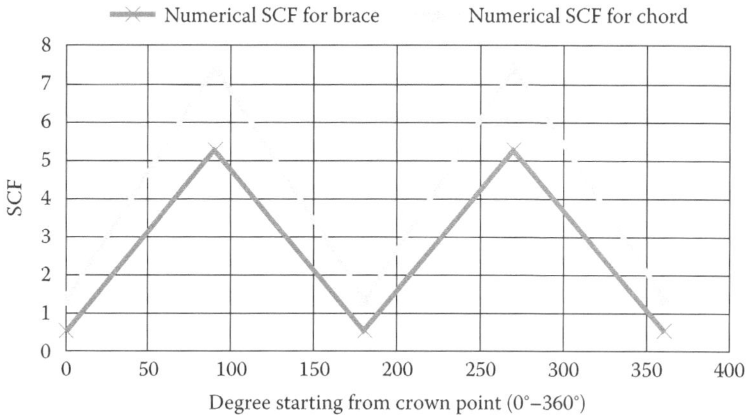

FIGURE 3.36 Stress concentration factor around weld of T joint under out-of-plane bending (numerical).

TABLE 3.11
SCF for T Joint under Out-of-Plane Bending (Parametric)

Member	Location	SCF
Brace	Saddle	6.21
Chord	Saddle	9.33

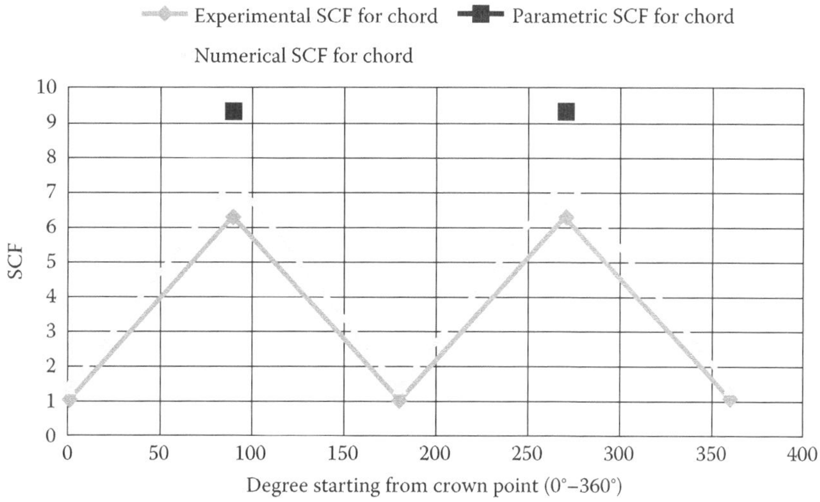

FIGURE 3.37 Comparison of SCF at T joint chord member.

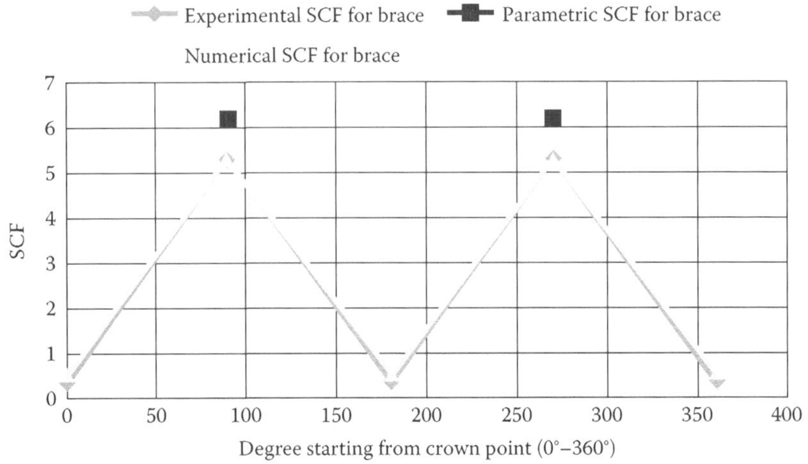

FIGURE 3.38 Comparison of SCF at T joint brace member.

TABLE 3.12
Constant k_1 for T Joint under Out-of-Plane Bending

Curve	$Log_{10}(k_1)$ When S is in MPa	m
Welded joints (WJ)	12.48	3 for $N < 10^7$
Welded joints (WJ)	16.13	5 for $N > 10^7$

experimental values of SCF agrees well. Table 3.12 gives the value of the constant k_1, required by Equation 3.118 as per the code (API RP2- WSD, 2001). Table 3.13 compares the estimate of number of cycles for T joint, under out-of-plane bending obtained from experimental investigations and parametric methods. Table 3.14 shows the fatigue life estimate for T joint under out-of-plane bending.

TABLE 3.13
Allowable Number of Cycles for T Joint under Out-of-Plane Bending Using Experimental and Parametrical SCF Values

Loading Condition	Location	Method	$б_{nom}$ (Nmm)	SCF	Δбo (Nmm)	Thickness Correction Δбo (Nmm)	N Cycle
OPB	Chord	Experimental	5	6.29	31.45	38.51,973	52,354,048
	saddle	Parametric	5	9.33	46.65	57.13,658	16,041,954
	Brace	Experimental	5	5.3	26.5	32.45,701	87,513,407
	saddle	Parametric	5	5.34	26.7	32.70,197	85,561,508

TABLE 3.14
Fatigue Life of T Joint under Out-of-Plane Bending

Loading	Method	*N*	No. of Cycle/Year	Fatigue Life (Year)
Out-of-plane bending	Experimental	52,354,048	500,000	104.70
	Parametric	16,041,954	500,000	32.08
	Experimental	87,513,407	500,000	175.02
	Parametric	85,561,508	500,000	171.12

It is observed that the fatigue life of the T joint under out-of-plane bending significantly increases from 32 years to 104.7 years when the experimental SCFs are used. This allows a higher service life for offshore structures. This is an important factor that influences uncertainty in reliability analysis when parametric equations are used to compute the fatigue life for T joints under out-of-plane bending in particular. It is observed during the experimental investigations that punching shear failure is followed by weld failure, circumferentially. The weld is seen to be intact until the yield point after which it resulted in instantaneous tearing of the weld near the chord. Figure 3.39 shows the failure mode of K joint under out-of-plane bending. The influence of the load on the chord is extended beyond the brace diameter region, which necessitates internal ring stiffeners. Maximum hot-spot stress is observed at the saddle point of the joint and the minimum is observed at the crown point of the weld. Table 3.15 shows the variation in SCF computed by experimental and numerical investigations and using parametric equations.

It is seen that the SCF arrived based on experimental investigations is lesser than that of the parametric, which intuits uncertainty encountered while using parametric equations. It is also seen that the fatigue life of the T joint under out-of-plane bending significantly increases from 32 years to 104.7 years when the experimental SCFs are used.

3.28 K JOINTS UNDER AXIAL LOADING

K joints are common to facilitate more intersections in the layout of the legs for braces, boat launcher supports, etc. Tubular K joints are experimentally investigated under axial loads to estimate the SCF and reliability. Table 3.16 shows the details of the K joints considered for investigations.

Figure 3.40 shows the experimental setup for a K joint under axial load.

Figure 3.41 shows the variation of SCF around the chord of a K joint under balanced axial load. while Figure 3.42 shows the variation around the chord for out-of-plane bending.

SCFs for the simple tubular K joint for balanced condition with the chord ends fixed and subjected to axial load are given in Table 3.17 (APIRP 2A). SCFs are also computed from the strain values measured experimentally. Table 3.18 shows the SCF values obtained experimentally for K joint under axial load. Figure 3.43 shows the variations of SCF values of K joint around chord under axial load.

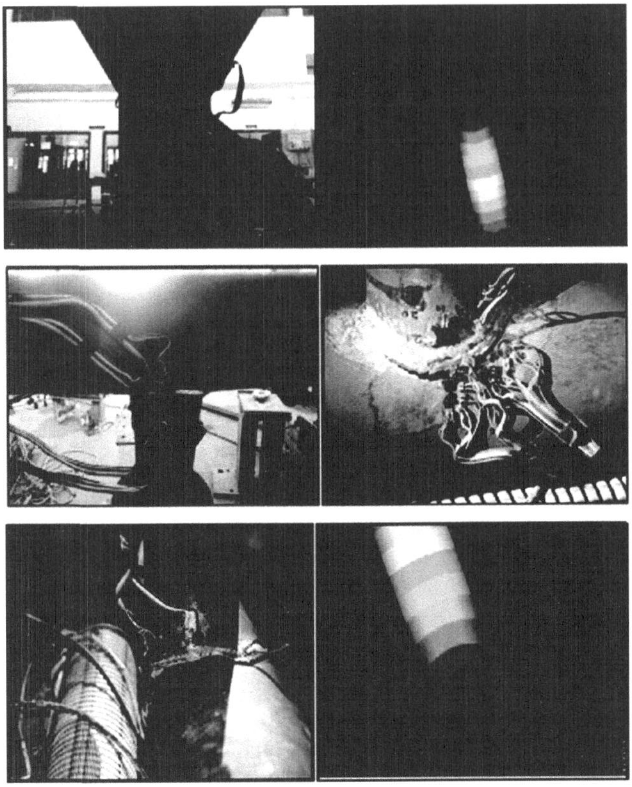

FIGURE 3.39 Failure mode of tubular T joint under out-of-plane bending.

TABLE 3.15
Variations in Stress Concentration Factor around T Joint Weld under Out-of-Plane Bending

Degree Starting from Crown Point (0°–360°)	Experimental		Parametric		Numerical	
	Chord	Brace	Chord	Brace	Chord	Brace
0	1.05	0.34	–	–	1.4	0.55
90	6.29	5.34	9.33	6.21	7.41	5.30
180	1.05	0.34	–	–	1.4	0.55
270	6.29	5.34	9.33	6.21	7.41	5.30
360	1.05	0.34	–	–	1.4	0.55

The average experimental SCF value for a tubular K joint under axial loading condition is 3.9. SCFs for K joint under axial load is also computed from parametric equations as shown in Table 3.19.

Table 3.20 gives the value of the constant k_1, required by Equation 3.118 as per the code (API RP2- WSD, 2001). Table 3.21 compares the estimate of number of

TABLE 3.16
K Joint under Axial Load

S. No	No.	Chord (mm)	Brace A (mm)	Brace B (mm)	β	τ	γ	α	⊖A	⊖B
1	K-1	168.3 × 7.11	114.3 × 6.02	114.3 × 6.02	0.7	0.85	11.84	4.46	45°	45°
2	K-2	168.3 × 7.11	114.3 × 6.02	114.3 × 6.02						
3	K-3	168.3 × 7.11	114.3 × 6.02	114.3 × 6.02						

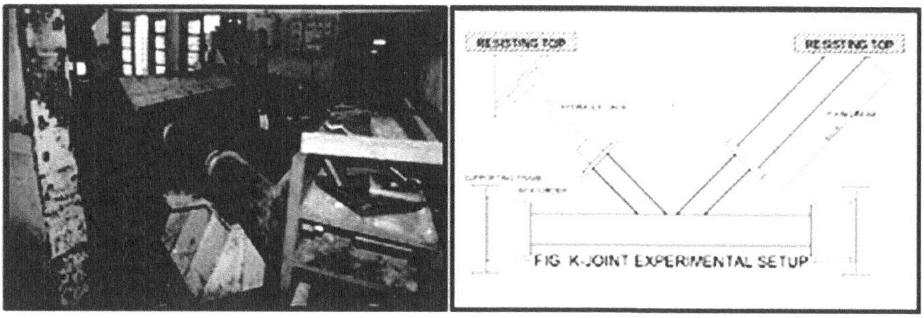

FIGURE 3.40 K joint under axial load.

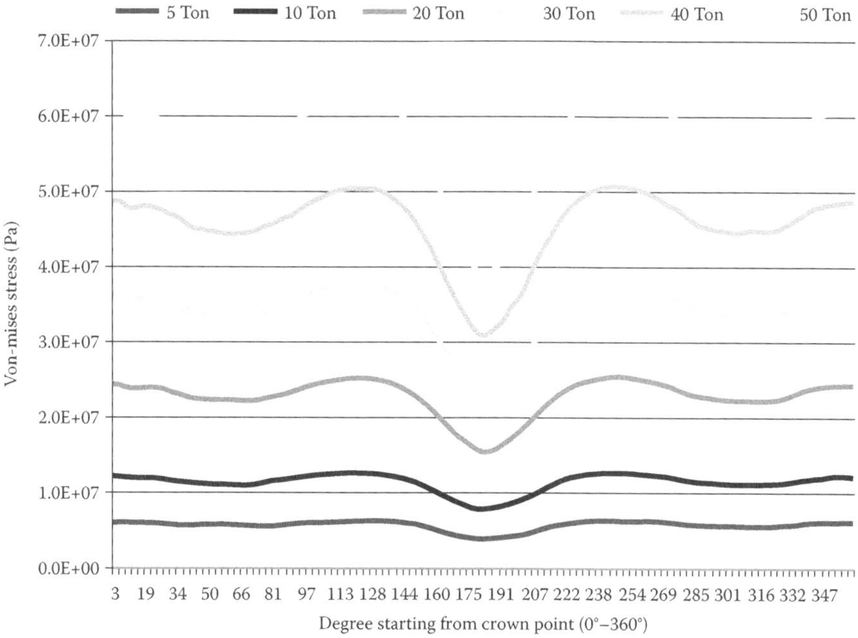

FIGURE 3.41 SCF variation around chord of K joint under balanced axial load.

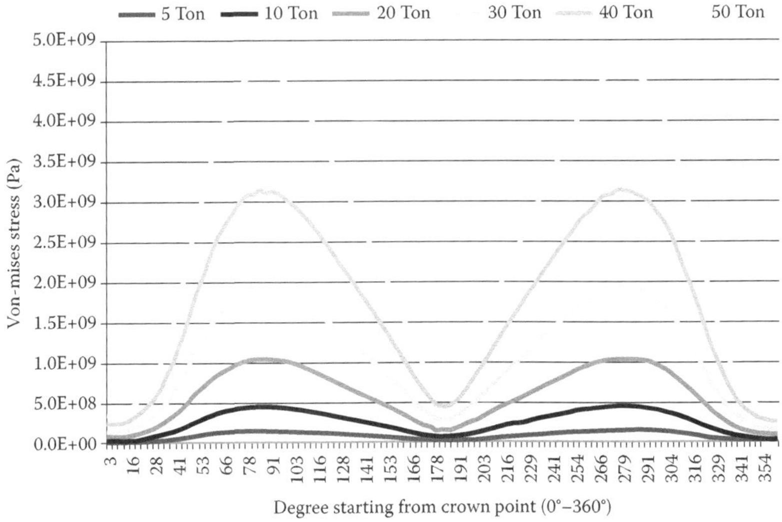

FIGURE 3.42 SCF variation around chord of K joint under out-of-plane bending.

TABLE 3.17
SCF Equation and Parameter Validation for K Joint under Axial Load

Location	Equation	Short Chord Correction
Chord	$\tau^{0.9}\gamma^{0.5}(0.67 - \beta^2 + 1.16\beta)\sin\theta(\sin\theta\, max/\sin\theta min)^{0.30}(\beta\, max/\beta\, min)^{0.30}(1.64 + 0.29\beta^{-0.38}\arctan(8\zeta))$(Equation 20)	None
Brace	$1 + (1.97 - 1.57\beta^{0.25})\tau^{-0.14}(\sin\theta)^{0.7}$(Equation 20) + $(\sin(\theta\, max + \theta\, min))^{1.8}(0.131 - 0.084\arctan(14\zeta + 4.2\beta))C\beta 1.5\gamma\, 0.5\tau^{-1.22}$	None

TABLE 3.18
SCF for K Joint under Axial Load (Experimental Investigations)

Member	Location	SCF
Brace A	Between 0° and 45° from crown point	3.2
Chord	Between 0° and 45° from crown point	3.9

cycles for a K joint under axial load, obtained from experimental investigations and parametric methods. Table 3.22 shows the fatigue life estimate for K joint under axial load.

It is observed that the fatigue life of the K joint under axial load increases from 341 to 439.27 years when the experimental SCF is used. This allows a higher service life for offshore structures. This is another important factor that influences

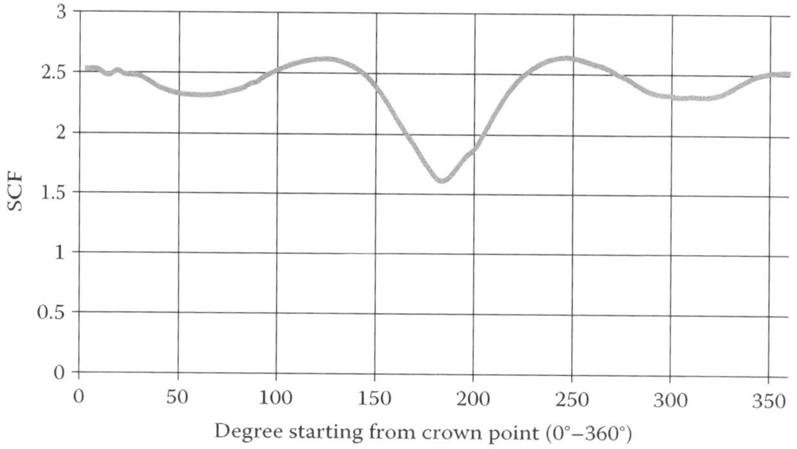

FIGURE 3.43 SCF of K joint under axial load (numerical analysis).

TABLE 3.19
SCF for K Joint under Axial Load (Parametric)

Member	SCF
Brace	2.86
Chord	4.24

TABLE 3.20
Constant k_1 for K Joint under Axial Load

Curve	$\log_{10}(k_1)$ When S is in MPa	m
Welded joints (WJ)	12.48	3 for $N < 10^7$
Welded joints (WJ)	16.13	5 for $N > 10^7$

TABLE 3.21
Allowable Number of Cycles for K Joint under Axial Load Using Experimental and Parametrical SCF Values

Loading	Member	Method	$б_{nom}$ (Nmm)	SCF	Δбo (Nmm)	Thickness Correction Δбo (Nmm)	N cycle
Axial load	Chord	Experimental	5	3.2	16	19.59,668	3.98E + 08
		Parametric	5	2.86	14.3	17.51,454	5.57E + 08
	Brace	Experimental	5	3.2	16	19.59,668	3.98E + 08
		Parametric	5	4.24	21.2	25.96,561	1.71E + 08

TABLE 3.22
Fatigue Life of K Joint under Axial Load

Loading	Member	Method	$N_{allowable}$	No. of Cycles/Year	Fatigue Life (Year)
Axial	Chord	Experimental	219,638,455	500,000	439.2,769,094
		Parametric	170,924,623	500,000	341.8,492,461
	Brace	Experimental	397,605,392	500,000	795.2,107,845
		Parametric	556,934,474	500,000	1113.868,947

TABLE 3.23
Variations in Stress Concentration Factor for K-Joint Weld under Axial Load

Member	Experimental SCF	Parametric SCF
Brace	3.2	2.86
Chord	3.9	4.24

uncertainty in reliability analysis when parametric equations are used to compute the fatigue life for K joints under axial loads, in particular. It is observed that K joint has a higher strength than that of T joint under axial loads. It is seen that the maximum stress is near the first crown (0–90°) and minimum at second crown (180°). Table 3.23 shows the variation in SCFs computed by experimental investigations and parametric equations.

It is seen that the SCF arrived based on experimental investigations is greater than that of the parametric, which suggests uncertainty encountered while using parametric equations.

4 Risk Assessment

4.1 INTRODUCTION

Oil and gas installations are part of and deeply intertwined with the wider world, a fact that becomes increasingly clear with the present globalization. The success of this industry widely depends on keeping the trust and goodwill of many stakeholders namely employees, suppliers, consumers, etc. It is obvious that such industries also perform operations under high risks. These operations include drilling, processing, and supplying oil to downstream units as a commercial end user product. It is also interesting to know that the fiscal growth/deficit index of a nation's economy is also controlled by the price of oil. It is therefore necessary to understand the risks involved in oil and gas industry in detail so that the efficiency of such industries can be studied with more confidence. In addition, offshore platforms and other installations are entangled by higher degree-of-uncertainties, which make the whole assembly under the coverage of the risk picture (Arsha Ayub, 2011; Bazzurro and Cornell, 1994a,b).

This industry involves some of the most ambitious engineering projects of the modern world, and is a prime source of revenue for many companies and countries. It also involves risks of major accidents, which have been demonstrated by disasters such as the explosion and fire on the UK production platform *Piper Alpha*, the capsizing of the Norwegian accommodation platform *Alexander Kielland*, the oil tanker *Torrey Canyon* grounded in the English Channel in 1967, the *Exxon Valdez* in 1989, the *Buncefield* fire in 2005 at the Hertfordshire Oil Storage Terminal, the pipeline rupture in the Usinsk area, Russia in 1994, etc. (Srinivasan Chandrasekaran, 2015b). Major accidents represent the ultimate, most disastrous way in which an asset of the offshore industry can be completely negated. Accidents cause death, suffering, pollution of the environment, and disruption of business. Being so dramatic, they attract attention from the news media and linger in the public memory, causing concern about safety (Bhattacharyya et al., 2010a,b). To understand the risk involved in the offshore industry, it is necessary to understand the basic fact that oil industries operate under the brackets of *acceptable risk* but never at zero risk. All the time, efforts are made only to bring down the level of risk to *as low as reasonably practical* (ALARP) but never to mitigate it completely (Bob Skelton, 1997; Bonvicini et al., 1998). One interesting question could be why such attempts are not made and not even planned. It is necessary to understand that risk mitigation is one of the most expensive schemes in any industrial investment. As oil companies are competitive in limiting production costs, investment is risk mitigation to make it to zero level is far too fanciful and will remain so until the means and methods for optimizing production costs do not become prevalent. A few terminologies are important:

Accident: A specific unplanned event or sequence of events that has undesirable consequences.

Hazard: A characteristic of the system/process plant that represents a potential for an accident causing damage to people, property, and (or) the environment.

Risk: A measure of potential economic loss or human injury in terms of the probability of the loss or injury occurring and the magnitude of the loss or injury if it occurs.

Explosion: A sudden release of energy accompanied by a blast wave.

Fire: A process of combustion, which is characterized by heat or smoke or flame or any combination of these.

4.2 QUANTIFIED RISK ASSESSMENT

Risk assessment studies are either qualitative or quantitative. Unfortunately, both are referred as QRA in the literature, which needs more detailed explanation to avoid this ambiguity. Qualitative risk assessment is a study or method to identify all possible hazards in the plant that have the potential to cause damage to personnel, property, and environment (David Brown and William, 2007). Various qualitative studies that are generally carried out in order to ensure (or to assess) safety about the process are namely: (i) safety audit; (ii) preliminary hazard analysis (PHA); (iii) hazard and operability (HaZop) study; and (iv) failure mode and effect analysis (FMEA). Quantitative risk assessment is a mathematical approach, which is commonly used to predict the risks of accidents and to give guidance on appropriate means of minimizing them (Dziubinski et al., 2006; Efthimia et al., 2012). It uses scientific methods and verifiable data. Different types of quantitative studies are namely: (i) Fault tree analysis (FTA); (ii) event tree analysis (ETA); and (iii) consequence analysis.

4.3 HAZARD IDENTIFICATION

Hazard analysis is intended for use in the preliminary phase of plant development for cases where past experience provides little or no insight into potential safety problems (Crawley et al., 2000; Engelhard et al., 1994). It does not preclude the need for further hazard assessment; instead it is a precursor to subsequent hazard analysis. *Hazard and Operability Study* identifies potential hazards and operability problems caused by deviations from the design intent of both new and existing process plants. *Failure Modes and Effects Analysis* is a tabulation of system/plant equipment, their failure modes, and each failure mode's effect on system/plant (Henselwood and Phillips, 2006). The technique is oriented toward equipment rather than process parameters. FMEA identifies single failure modes that either directly result in or contribute significantly to an important accident. This evolves risk priority number (RPN), which is a useful tool to identify weak links in the preliminary design of the system, both in mechanical and process parameters.

4.4 HAZARD AND OPERABILITY

Hazard and Operability (HaZop) study is a structured and systematic examination of a planned or existing process or operation in order to identify and evaluate problems that may represent risks to personnel or equipment, or prevent efficient operation (IEC 61882, 2000; IS 15656, 2000). The most comprehensive qualitative hazard analysis method is the HaZop study approach, which has received wide acceptance by both the process industries and the regulatory authorities. HaZop is a tool often used for investigating accidental happenings across the process industries to find what went wrong. It is one of the best tools to postmortem an accident scenario, but can also be used to reduce risk level in the process plant. Risk in the top side of offshore installations are generally analyzed with HaZop to assess the extent of risk involved in terms of human and asset safety. It gives brief information about the cause of the accident, its consequence, and necessary action taken to prevent such accidents in the future. It uses two types of keywords, primary and secondary to identify various hazards that are already present or perceived. The HaZop study is intended to examine the overall process through brain storming sessions and search for operating deviations and process interactions (Jan Erik, 2007). Such intrinsic examination shall lead to hazardous situations or operability problems, which may include: (i) safety and occupational health hazards to personnel; (ii) damage to equipment/asset/environment; (iii) operability/maintainability problems; (iv) plant nonavailability/limitation and lack of product quality/production loss; (v) environmental emissions; (vi) demolition/decommissioning/abandonment reviews; and (vii) construction and commissioning hazards (Khan and Abbasi, 1999).

4.4.1 Applicability

The HaZop technique is used to identify the potential hazards associated with complex chemical processes including any system/facilities, existing or new, which has a process flow sequence, batch, continuous or sequential operations (Kyriakdis, 2003). HaZop studies shall also be used to examine hazards and potential problems associated with different operating states of a given system. For example, start-up, normal operation, normal shutdown, and emergency shutdown problems can be analyzed with HaZop studies (Nivolianitou et al., 2006). Applicability of HaZop studies during the various stages of a project is described as below.

New facilities or enhancements or modifications: HaZop studies can be applied to all process designs involving new facilities, facility enhancements or modifications, including package units and utility systems. Systematic HaZop studies, conducted at the appropriate stages of the project ensure identification of any potential hazards and operability problems. It also guides initiation of early actions to mitigate such problems, which will help in smooth start up and safe operation of the process unit. This technique is normally applied when the design is essentially complete and the piping and instrumentation diagram (P&ID) have been developed with all essential information. HaZop is conducted after finalization of front end engineering design (FEED), which will improve the process safety and help in reducing high risk recommendations at a later stage.

Existing facilities: HaZop is considered before implementing any changes to existing facilities that would affect the safety or operability of a system or have environmental impacts. This technique can be applied to review existing plant P&ID's or procedures to identify potential hazards and operability problems which have not already become evident from operating experience or to review nonstandard operation or procedures not previously covered. Modifications involving changes to P&ID's, changes in operation and the associated safety systems can also be subjected to HaZop studies. Basic assumptions made in HaZop studies are namely: (i) process design and documentation under the study is basically sound with respect to engineering application of relevant codes and standards; and (ii) when a process is operating within its operating envelope in a stable condition as per the design, there is no potential for hazards (OSID, 2002; Papazoglou et al., 2003; Pate Cornell, 1994).

4.5 HAZOP STUDY PROCESS

Figure 4.1 shows the algorithm for HaZop study process.

4.5.1 Node Identification

The system under consideration will be divided into various sections containing process lines and/or equipment. The HaZop study progresses through the plant, node-by-node. Selection of the node sizes and the route, through the plant is analyzed, are decided by the facilitator, before the start of the study. Guidelines for the facilitator to follow are based on changing nodes at the location of major equipment or where a change in a critical parameter like flow, pressure, or temperature is likely to occur. The main objective is to focus on appropriately sized sections of the system to clearly identify all the design and operational issues. Nodes should be described in terms of: (i) brief description of the operational parameters; (ii) operating and design conditions; (iii) method of operation and maintenance; and (iv) specific requirements for operational intervention. Before the study meeting, the facilitator should record the data on a node detail form, which can be reviewed by the HaZop study team during the HaZop study.

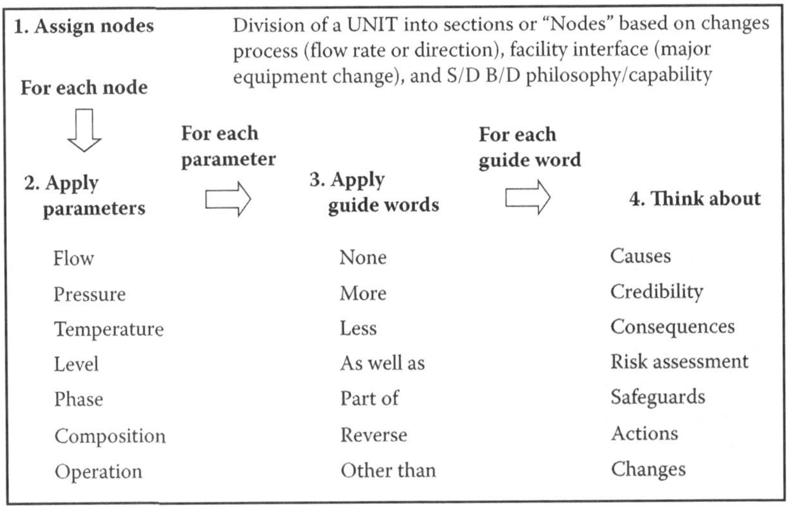

FIGURE 4.1 HaZop study process.

4.5.2 Operating Modes

All potential operating modes should be identified and included in the study. Each mode shall impose different maximum limits in the operating envelope. It is also usual to identify that some nodes are used for different purposes under the various operating modes of the unit. Process conditions may also change significantly during start-up or upset conditions and therefore should be clearly mentioned in the study. Maintenance isolation of individual or a group of equipment sections may have significant effects on other parts of the process by changing the flow rates or temperatures/pressures/phases within the surrounding sections. It is important to take care of such intentional deviations, which may be caused during operational maintenance.

4.5.3 Lifecycle Changes

There are often changes to material integrity throughout its lifecycle, caused by operation outside design, and changes in working fluid. Information on the history of a component may be found in the records of operational data. This information may be available to the HaZop study team as background information.

4.6 PARAMETERS FOR HaZop STUDY

There are seven basic parameters namely: *flow, pressure, temperature, level, phase, composition,* and *operation* which are vital for the HaZop study. The first three are usually regarded as the main parameters; additional parameters may be selected for some of the nodes in a study as appropriate (Srinivasan Chandrasekaran, 2010a,b, 2011a; Srinivasan Chandrasekaran and Kiran, 2015; Srinivasan Chandrasekaran and Madhuri, 2015). It is important to note that all parameters are not applicable to each node. Flow is usually applied as the first parameter as it is the easiest parameter to visualize and also to determine the effects of its consequence. Pressure is also one of the vital parameters as it leads to many of the most common hazards in process industries like offshore plants. Areas causing common pressure-related problems are namely: HP/LP interfaces, isolation philosophy, vessel protection, and compressor/pump maximum outputs. High and low temperatures may exceed the design parameters of the equipment or cause hazards to the personnel and are therefore considered as vital. Phase changes may also occur due to freezing, boiling, and gas–oil ratio (GOR) changes (Srinivasan Chandrasekaran, 2011b–e). Low temperature may lead to waxing or hydration, causing blockages. Level is usually relevant inside vessels only where level control issues may arise, including flooding or gas blow-by. Many level hazards overlap with other deviations, for example, “more flow” gives “more level.” In such cases, level may be easier for the team to visualize, and for vessels the facilitator may decide to use level as the first parameter, rather than flow. In addition, a few parameters, which will often cover phase changes caused by the pressure/temperature envelope are also used in the study. It is sometimes relevant for a plant where phase changes can be caused by extreme operating conditions, for example, solids, flashing, or slugging in flow lines. On hydrocarbon plants, parameter composition normally covers issues where there are fewer or more components than expected, for example, more water or less hydrocarbon. It also includes the possibility of contamination from other process streams, for example, hydrocarbon in a cooling water circuit, or the presence of unwanted components, for example, H_2S or radioactive material. Lack of a component is also often a hazard, for example, corrosion inhibitor. Composition deviations may also be revealed under other parameters by application of *as well as* and *part of* to flow. Operation issues may be dealt with at any time during the study, usually under the guide words *other than*. It is however often a very useful option to discuss them together in a separate "operation" parameter at the end of each node once all the process parameters have been discussed. Operation may be used to cover all namely: isolation philosophy, failed operation or maintenance, start-up, shut down, maintenance access, and normal operation or maintenance.

4.6.1 Guide Words

Guide words are simple words or phrases used to qualify or quantify the intention and associated parameters in order to suggest deviations. There are seven standard guide words: *no, more, less, as well as, part of, reverse,* and *other than,* but not all are applicable to each parameter, for example, "no" or "part of" do not apply to the parameter of temperature. *Other than* is very popular as a "catch all" guide word at the end of each parameter. There is a significant amount of overlap in parameter guide word combinations. If some deviations appear to be already covered, for example, "*more flow*" may lead to "*more level*," then the decision to move on should be made specifically for each guide word and not by assuming that a whole parameter is already covered. A minute saying "*see more flow*" should be added in the "*more level*" section of the worksheet. Care is, however, required, as it is possible to overlook possible hazards. For example, "*more flow*" resulting in "*more level*" may not cause additional level problems, but may result in velocity issues. Following are the basic guide words used for a HaZop study:

- *No:* Negation of the design intent (e.g., no flow when there should be; no pressure there should be)
- *Less:* Less of a physical property than there should be, quantitative decrease (e.g., lower flow rate than there should be)
- *More:* More of physical property than there should be, quantitative increase (higher flow, temperature, pressure, viscosity, etc., also actions-heat and reaction)
- *Part of:* Composition of the system (stream) is different than it should be, qualitative decrease (e.g., less of one component)
- *As well as:* More components present than there should be, qualitative increase (e.g., extra phase or impurities present)
- *Reverse:* Logical opposite of the design intent (e.g., reverse flow)
- *Other than:* Complete substitution (e.g., transfer of a material other than the material intended; transfer of a material to location other than intended)

4.6.2 Deviations

The combination of parameters and guide words, in sequence, will identify all the deviations (no flow, more temperature, etc.). There may be significant overlap between the deviations considered (no flow may have the same effect as more pressure). These will be identified and excluded during the team discussion. Deviations are the key to the HAZOP study process. They are departures from the design intent of the process and are identified by the systematic application of the appropriate parameter/guide word combinations. In some cases the parameter and guide word when combined make a well understood deviation, for example, *less flow*. In other cases explanations may need to be developed for the deviation, for example, *as well as flow* may more easily be described as "additional component," and *part of flow* arises when an element of the flow composition is missing. Table 4.1 shows list of common guide words and their usage.

4.6.3 Causes

The team will brainstorm to find all the potential causes of the deviation. There is frequently more than one cause, for example, *no flow* would be caused by a blockage, but there may be several different causes of that blockage. All potential causes should be identified and discussed as the consequences and actions may be different. If only the generic cause is covered, methods of prevention and operational issues will be missed. It is not sufficient to refer to generic causes, for example, blockage, since the method of prevention and operational issues cannot be adequately defined. Causes have to be credible, for example, impact damage from the explosion of the neighboring vessel is only possible if the vessel contains flammable material or is at great pressure. If incredible or highly unlikely causes are considered by the team this might lead to expensive and unnecessary modifications to the plant. The three major categories of causes in decreasing probability are namely: human error, equipment failure, and external events.

TABLE 4.1
Parameter and Guide Word Combinations for HaZop Study

		Guide Words						
		None	More	Less	As Well as	Part of	Reverse	Other than
Parameters	Flow	+	+	+	+	+	+	O
	Pressure	X	+	+	X	X	X	O
	Temperature	X	+	+	X	X	X	O
	Level	+	+	+	X	X	X	O
	Phase	+	+	+	X	X	X	O
	Composition	+	+	+	X	X	X	O
	Operation	+	+	+	+	+	+	+

Key: X—not used; +—used; O—possible.

4.6.4 Consequences

Potential consequences for each deviation are discussed and assessed within the limits of the information available and the expertise of the team. There may be several consequences involving escalation to other pieces of equipment. Having identified the credible causes of the deviation, discussion should move on to analyze and assess the significance of the consequences. The role of the HAZOP study is to determine the net effect of the potential consequences and the mitigating effect of the safeguards. This may be done by ranking the consequences in terms of likelihood and severity to determine risk. Risk ranking should not take into consideration any safeguards.

Likelihood: An assessment of the approximate frequency of most events and from this the likelihood of the deviation under consideration, within ranges from never heard of (range 0) to often (range 5).

Severity (consequence): As with likelihood, the severity or consequences may also be assessed in terms of approximate magnitude. Again, a five-category system is used, from no injury/no damage (range 0) to multiple fatality, extensive damage, or massive leak (range 5). In the case of several consequences, the "worst case" should be selected.

Risk: The combination of likelihood and severity will lead to an approximation of the risk involved for each hazard. The risk level is then applied to the importance of any actions associated with the deviation.

Where assessment of the risk cannot be done by the team, and there is residual concern regarding the severity of the consequence, further in-depth analysis may be recommended.

4.6.5 Safeguards

Most hazards will be protected against by controls, a combination of hardware, instrumentation, and operating practices. Protection may be provided by a combination of the following

- The intrinsic mechanical strength of the plant
- Process control responding to the deviation
- Alarms to alert operators to manually intervene in the process
- Plant trips and shutdowns
- Relief systems being activated

It should be recognized that protective systems have a finite failure rate. The team should consider how probable a failure would be, and also the potential consequences of any failure. A single high-level alarm may be acceptable on a water-storage tank but would be unacceptable on the suction scrubber of an expensive compressor. The protective elements provided for the facility may fail for the following reasons:

Human failure: Operators may fail to prevent a hazard by inappropriate response to an alarm due to wrong procedure/instruction, inadequate time, or inadequate data for the diagnosis of the problem. The manning philosophy and associated reaction time should always be analyzed when hazards are mitigated by alarm and operator intervention. Operator action should normally follow the activation of a pre-alarm, giving time to intervene before major executive action by the control system is required.

Patent failure (hardware): A patent failure is one where the failure is immediately apparent (self-revealing), for example, a normally open control valve fails shut and the process flow stops. These types of failure are normally easily identified as causes of deviations.

Latent failure (hardware): A latent failure occurs when a fault is not immediately apparent. It is usually caused by a protective device (e.g., a high-level alarm or switch failure) which will not reveal itself until a demand is made on it by the failure of the normal control function. This is a common problem but must not be confused with "double jeopardy" which is defined as two *simultaneous* and *independent* failures. In most HAZOP studies, double jeopardy is assumed to have a very low probability and is normally excluded. The discussion from the study is recorded on log sheets. Information is recorded in columns on the log sheets for each identified *node* as shown in Table 4.2.

The study is performed by a team of experts from different backgrounds working under the guidance of a leader who is experienced in use of the HaZop study. The following steps are carried out for the study:

- Identify a section of plant on the piping and Instrumentation diagram (P&ID).
- Define the design intent and normal operation conditions of the section.
- Identify a deviation from design intent or operating conditions by applying a system of guide words.
- Identify possible causes for, and consequences of, the deviation. A deviation can be considered meaningful if it has credible cause and can result in harmful consequences.
- For a meaningful deviation, decide what action, if any is necessary.
- Record the discussion and action.

The method identifies the causes of a deviation from the design intent by application of the guide words. The ultimate consequences are then identified and recorded without reference to the safeguards, which are in place (examples of safeguards are plant design, control systems, and procedures). The consequences are then compared with the safeguards and the HaZop team then decides whether the current safeguards are adequate. If the team considers them inadequate, then a recommendation is made to consider or add further safeguards. The HaZop study is based on the piping and instrumentation diagrams (P&ID), the following documents are used during the HaZop sessions:

- Piping and instrumentation diagrams (P&ID)
- Flowcharts

TABLE 4.2
Typical HaZop Record Sheet

Guide Word	Deviation	Causes	Consequences	Existing Safeguards	Risk Ranking			Risk Category	Recommendations		Comments
					S	L	R	SA/OP/EN/OH	No		

S—severity; L—likelihood; R—risk ranking;
SA—safety; OP—operational; EN—environmental;
OH—occupational health.

- Process description
- Operating manual
- Equipment/instrument specification
- Supplier manual
- Accident records
- Material safety data sheet (MSDS) of chemicals

4.7 HaZop: ADVANTAGES AND LIMITATIONS

A HaZop study has many advantages and of course a few limitations. HaZop is one of the best and systematic ways of identifying hazard, which has potential to mature to a risk in an offshore plant. The following are typical advantages of a well-organized HaZop study:

- Systematic examination
- Multidisciplinary study
- Utilizes operational experience
- Covers safety as well as operational aspects
- Solutions to the problems identified may be indicated
- Considers operational procedures
- Covers human errors
- Study led by independent person
- Results are recorded
- Assumptions for doing HaZop
- Double failures in the plant would not be considered at one time
- Failure of safeguards will not be considered

4.7.1 Limitations of HaZop

- Accuracy of drawings and data used as a basis for the study
- Experience and skills of the HaZop team leader
- Technical skills and insights of the team
- Ability of the team to use the HaZop approach as an aid to identify deviations, causes, and consequences
- Ability of the team to maintain a sense of proportion, especially when assessing the severity of the potential consequences

4.7.2 Example Illustration

The HaZop study is illustrated through an example. Figure 4.2 shows the typical flow diagram of a process plant. The dosing tank receives chemical dosage at a particular concentration from another supply line, whose segment is isolated from this study. As we understand, HaZop is carried out segment-wise and hence this example is focused on the chosen segment alone. The chosen dosage is required to be supplied to vessel V1, which is done through the pipeline housing strainers and pressure valves. To compensate for the pressure loss, the dosage chemical is pumped at high

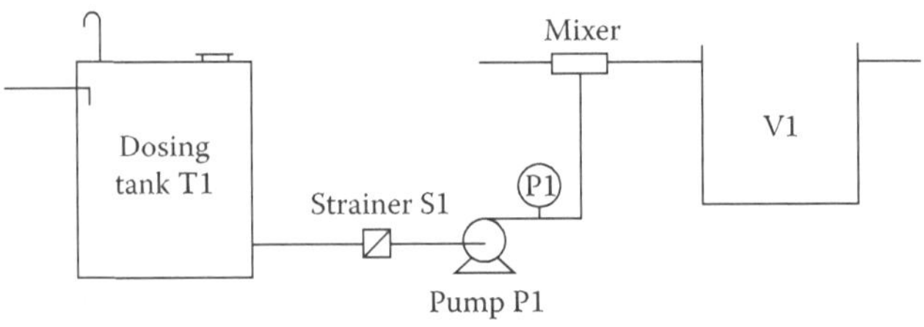

FIGURE 4.2 Isolated segment of process line for HaZop study.

pressure using pump P1. As the vessel V1 is expected to be free from any floating or dissolved impurities since this may affect the operational efficiency of the down line of the process, strainer S1 is also included in the line. Table 4.3 shows the typical HaZop sheet, indicating one of the possible hazards that may arise in the chosen example problem.

4.8 LOGICAL RISK ANALYSIS

Frank and Morgan developed a method of logical risk analysis in 1979 (Frank and Morgan 1979). The proposed method is a systematic approach to financing risk toward risk reduction. A very classical example of this is as follows, that is, if one wants to reduce the risk in life one generally goes in for health insurance. So, insurances are nothing but financing risk toward one's risk reduction. This method involves six steps of risk analysis as follows:

- Step 1: Compute the risk index for each department
- Step 2: Determine relative risk for each department
- Step 3: Compute present risk index for each department
- Step 4: Compute composite exposure for each department, which is the financial aspect of converting risk into money
- Step 5: Compute composite risk for each department
- Step 6: Rank all the departments, relative to each other based on composite score

TABLE 4.3
HaZop Sheet for the Example Problem

Deviation	Cause	Consequence
FLOW/NO	Potential causes that would result in deviation—strainer S1 blockage due to impurities in the dosing tank T1 might cause no flow	• Consequences that arise from the effect of deviation—loss of dosing results in incomplete separation in V1 • Cavitations in pump P1. This may result in possible damage if prolonged

Let us take an example to explain this in detail. Let us consider a process industry, which aims for risk assessment. The plant is categorized into a convenient number of departments which is enclosed in the industrial umbrella. For example, research and development, production, maintenance, inventory, etc. It is assumed that the data of categorizing the plant into different departments already exist in industrial management.

4.8.1 Risk Index

Each department has an inherent risk level, which is to be first identified. This can be easily done by evaluating hazards present and the control measures available. This is also called as the first level of risk assessment. A checklist is developed by Morgan with a hazard score and control score assigned to each criteria of the checklist as shown in Table 4.4. A similar checklist can be prepared for individual industries assigned with hazard score and control score corresponding to the industry. A detailed checklist for each department needs to be prepared assigning a hazard score and a control score with the established Morgan checklist. The checklist gives different rating points for different hazard groups. For example, if we have a hazard group of fire and explosion potential and there are large inventory of flammables available, the rating point is given as 2. If the flammable chemicals are processed above the flash point, the rating point is given as 1. Based on the complexity of the process, the maximum score is given as 8. If precise reactant addition and control are needed then the rating point is given as 2. If it is difficult for start up or shut down, the complexity of the process rating is given as 1. Based on the stability of the process, if it is severity of an uncontrolled situation, the rating point for stability of the process is given as 3. Similarly, if any obnoxious gases are present or stored under pressure, the stability of process is given as 1. Similarly, it can be observed that the rating points for operating pressure involved personal and environmental hazard potential available and high-temperature scenario. So, based on different kinds of hazard groups, a rating point can be easily fixed. These values are subjective and a similar checklist can also be prepared based on experience.

Once the checklist is prepared, control groups are observed and the rating point for different control activities present in the plant are given. For example, if there are automatic sprinkler systems capable of meeting the demands available in the plant for fire protection, the rating point is given as 4. If the fire protection system is inspected in this state, with regular frequency, the fire protection rating point is given as 10 which are all control group points. Different rating points can be given based on the electrical integrity, inserted safety devices, deep piping, presence of ventilation, and kinds of control measure that is available in the plant for safety, fire protection, etc. These rating points for different subgroups are given in Table 4.4. For example, Morgan suggested a checklist where the control group is divided into 1, 2, 3, 4, and 5 divisions like fire protection, electrical integrity, safety devices inserting and dip, piping, and ventilation. So, if one has an experience of auditing a similar plant earlier, one can also prepare a similar checklist and have more than five groups as well. The idea is to try to give different dating point for different existing control mechanisms.

TABLE 4.4
Parameters Useful for Risk Analysis

Rating Points	Control Group and Control (Group Control Factor in Parentheses)
	Fire Protection (10)
4	Automatic sprinkler system capable of meeting demands
2	Supervisors and operators knowledgeable of installed fire protection systems and trained in proper response to fire
1	Adequate distribution of fire extinguishers
1	Fire protection system inspected and tested with regular frequency
1	Building and equipment provided with capability to isolate and control fire
1	Special fire detection and protection provided where indicated
	Electrical Integrity (8)
3	Electrical equipment installed to meet National Electrical Code (NEC) area classification
1	Electrical switches labeled to identify equipment served
1	Integrity of installed electrical equipment maintained
1	Class I, division 2 installations provided with sealed devices. Explosion proof equipment provided or purged reliably and good electrical isolation between hazardous and nonhazardous areas
1	All electrical equipment capable of being locked out
1	Disconnects provided, identified, inspected, and tested regularly
1	Lighting securely installed and facilities properly grounded
	Safety Devices (7)
3	Relief devices provided and relieving it to safe area
2	Confidence that interlocks and alarms are operable. Operating instructions are complete and current, and department has continuing training and/or retraining program
1	Safety devices are properly selected to match application
1	Critical safety devices identified and included in regular testing program
1	Fail-safe instrumentation provided
	Inerting and Dip Piping (5)
2	Vessels handling flammables provided with dip pipes
2	Vessels handling flammables provided with reliable "inerting" system
2	Effectiveness of inciting assured by regular inspection and testing
1	lnerting instruction provided and understood
1	Inciting system designed to cover routine and emergency start-up
1	Equipment grounding visible and tested regularly
1	Friction hot spots identified and monitored
	Ventilation/Open Construction (4)
3	No flammables exist or open air construction is provided
2	Local ventilation provided to prevent unsafe levels of flammable, toxic, or obnoxious vapors
2	Provision made for containing and controlling large spills and leaks of hazardous materials
1	Building design provides for natural ventilation to prevent accumulation of dangerous vapors

(Continued)

TABLE 4.4 (*Continued*)
Parameters Useful for Risk Analysis

Rating Points	Control Group and Control (Group Control Factor in Parentheses)
1	Sumps, pits, etc., nonexistent or else properly ventilated or monitored
1	Equipment entry prohibited until safe atmosphere assured
	Accessibility and/or Separation (2)
2	Critical shutdown devices and/or switches visible and accessible
	Adjacent operations or services protected from exposure resulting from incident in concerned facility
2	Operating personnel protected from hazards by location
1	Orderly spacing of equipment and materials within the concerned facility
1	Adjacent operations offer no hazard or exposure
1	Hazardous operations within facility well-isolated

Rating Points	Hazard Group and Hazard (Group Hazard Factor in Parentheses)
	Fire/Explosion Potential (10)
2	Large inventory of flammables
2	Flammables generally distributed in the department rather than localized
2	Flammables normally in vapor phase rather than liquid phase
2	Systems opened routinely, allowing flammable/air mix, versus a totally closed system
1	Flammables having low flash points and high sensitivities
1	Flammables heated and processed above flash point
	Complexity of Process (8)
2	Need for precise reactant addition and control
2	Considerable instrumentation requiring special operator understanding
2	Troubleshooting by supervisor rather than operator
1	Large number of operations and/or equipment monitored by one operator
1	Complex layout of equipment and many control stations
1	Difficult to start-up or shut-down operations
1	Many critical operations to be maintained
	Stability of Process (7)
3	Severity of uncontrolled situation
2	Materials that are sensitive to air, shock, heat, water, or other natural contaminants in the process
2	Potential exists for uncontrolled reactions
1	Raw materials and finished goods that require special storage attention
1	Intermediates that are thermally unstable
1	Obnoxious gases present or stored under pressure
	Operating Pressure Involved (6)
3	Process pressure in excess of 110 lb/in^2 (gauge), or
2	Process pressure above atmosphere but <110 lb/ins (gauge), or

(*Continued*)

TABLE 4.4 (*Continued*)
Parameters Useful for Risk Analysis

Rating Points	Hazard Group and Hazard (Group Hazard Factor in Parentheses)
1	Process pressure ranges from vacuum to atmospheric
3	Pressures are process rather than utility related
2	High pressure situations are in operator-frequented areas
1	Excessive sight glass application
1	Nonmetallic materials of construction in pressure service
	Personnel/Environmental Hazard Potential (4)
3	Exposure to process materials pose high potential for severe burns or severe health risks
2	Process materials corrosive to equipment
2	Potential for excursion above threshold limit value (TLV)
1	Spills and/or fumes have high impact on equipment, people, or services
1	High noise levels make communication difficult
	High Temperatures (2)
1	Equipment temperatures exist in <100°C range (low), or
2	Equipment temperatures exist in 100 < 170°C range
3	Equipment temperatures exist in 170 < 230°C range (3501 h/in^2 [gauge] steam)
2	High-temperature situations are in operator-frequented area
2	Overflows and/or leaks are fairly common
2	Heat stress possibilities from nature of work or ambient air

The Morgan hazard checklist has six groups of hazards (Frank and Morgan, 1979). There are points associated with each group, as seen in two different tables. These points given under each group are summed up for the hazards applied within that group. Now, hazards score for a given group is sum multiplied by the hazard weightage for that particular operation. In this manner, one can easily identify or determine the hazards score for each department, which is nothing but the sum of scores computed for each of the six groups. Similarly, it is done for the control score establishment as well. Control score for a department is again the sum of scores of each of the six groups, as seen in the table. Now, it is very simple to estimate risk index.

$$\text{Risk index} = \text{Control score} - \text{Hazard score}$$

It should be noted that the risk index can be a negative number as well. For example, if the control score for the department is very poor and the department has a very high hazard potential, the risk index can become a negative number. Any department, which has a positive risk index, is comparatively safe with respect to the other department which has a negative risk index. So, the risk index score may be positive or negative.

4.8.2 Relative Risk

The relative risk is an estimate of risk between the departments. The aim is to rank the departments and not the individual hazards of each department. This is because, the department with highest risk index (highest positive value) is not likely to need much reduction in hazards. It is about financing risk reduction. A high-risk index means that the control measures are very effective. Those departments will need less funds than the other departments to mitigate, eliminate, or reduce hazards. Once the relative risk is estimated, the best department risk score is used as a base reference. The department which has highest positive risk score is considered as the best department. All scores are then adjusted relative to the score of the best department, by subtracting the risk score of best department from all other risk scores. This adjustment will make the relative risk of the best department zero.

$$\text{Relative risk index} = \text{Risk index of individual dept} - \text{Maximum of risk index of all depts}$$

Now the percentage of risk index is worked out. The area of interest is the percentage of risk index of each department. This indicates relative contribution of each department to the total risk of the plant. This calculation gives us the department which is contributing the maximum total risk of the complete plant. Then a percentage contribution of each department to the overall risk scenario of the whole plant is compared. The relative risk of each department is converted to a percentage of all the risk by a simple procedure. The total risk of all departments is a sum of the absolute value of relative risk of each department (as relative risk can be a negative number as well) and subsequently the percentage is estimated.

Risk evaluated as above is then converted to financial terms, called composite exposure. The estimated risk is now converted to a financial value, because we are now looking at financing risk reduction. This estimates the financial value of risk of every department. Composite exposure is the sum of monetary value of three components of every department. Every department has a property value and every department can contribute to a loss, if business in the department is interrupted. Also, every department has different classes of personnel and the exposure level of these personnel can also be different. Now a simple exposure value is associated to each one of them separately. Morgan suggested a very interesting process by which this can be done. The property value is simply estimated by replacement cost of all material and equipment at risk in the department. For example, if a department has some material and equipment at risk, that value of the material and equipment can be simply called the property value of the department in terms of money. Business interruption, as explained earlier, is computed as the product of the unit cost of goods multiplied by the department production per year multiplied by expected percentage of capacity. So, if the business in the department is interrupted, this would be the total loss the department will contribute to the overall loss of the whole plant. Personnel exposure is simply the product of total number of people in the department during the most populated shift, multiplied by the monetary value of each person. The monetary value of each person involved in the department may be different, and the populated shift can have a mixture of such kind of people. Hence, the product of these two, evaluated carefully, will give personnel exposure. Composite risk is estimated. For each department, this will be a simple product of composite exposure multiplied by percentage risk index of every department. This represents the value of relative risk of every department. Next the final ranking is conducted. The main objective of Morgan's method is to rank the departments depending upon the risk involvement in the production process. This is the final step in the process where the departments are ranked based on the composite risk of the department. The composite risk is introduced as this helps the risk managers to decide the level of funds required by each department. Hence, the ranking of a department is mandatory, based on which a risk manager should be able to decide what level of funding should be given to each department for risk reduction or risk mitigation. Therefore, it is inevitable that the department should be ranked from the highest composite score to the lowest. The lowest will be zero, which is considered as the reference department with the maximum positive risk score. This whole process of Morgan risk assessment can be explained by a simple example of a process industry. Table 4.4 shows various parameters that are useful to obtain the risk ranking of the process plant. A typical table, indicating other factors can also be developed by the user, which can be subsequently used in the risk analysis (Ramamurthy, 2011).

4.8.3 Example Problem for Risk Analysis

Let us consider an example of process industry that has seven departments, namely, A, B, C, D, E, and F. For each department the following data is given below.

Dept.	Hazard Score (HS)	Control Score (CS)	Property Value (PV) (Lakhs)	Interception Cost (IC) (Lakhs)	Personnel Value (PC) (Lakhs)
A	185.00	200.00	3000.00	1500.00	444.00
B	65.00	149.00	600.00	1100.00	450.00
C	162.00	88.00	1700.00	450.00	1100.00
D	200.00	150.00	1000.00	450.00	600.00
E	100.00	60.00	3222.00	3200.00	2200.00
F	150.00	141.00	1000.00	1200.00	400.00
G	200.00	250.00	2000.00	1000.00	100.00

Solution:

Dept.	Risk Index (HS–CS)	Relative Risk Index	% Risk Index (%RI)	Composite Exposure (CE = PV + IC + PC)	Composite Risk (%RI × CE)	Rank
A	15.00	−69.00	11.27	4944.00	55741.18	3
B	84.00	0.00	0.00	2150.00	0.00	7
C	−74.00	−158.00	25.82	3250.00	83905.23	2
D	−50.00	−134.00	21.90	2050.00	44885.62	4
E	−40.00	−124.00	20.26	8622.00	174694.12	1
F	−9.00	−93.00	15.20	2600.00	39509.80	5
G	50.00	−34.00	5.56	3100.00	17222.22	6

4.9 FAILURE MODE AND EFFECT ANALYSIS

Failure mode and effect analysis (FMEA) is an alternative method of hazard identification. It considers all possible outcomes from all failure modes or deviations of complex mechanical and electrical systems and can be applied at different levels of complexity. FMEA systematically identifies consequences of component level failure and determines significance of each failure with regard to the system's overall performance. This method is primarily used to study material and equipment failure of the system tracing each failure mode within the system. For example, we identify a mechanical or electrical system, in that system we identify multiple failure modes, and for each failure mode we find its effect on the overall performance. When there are multiple failure modes, we then try to identify the sequence of this failure, so that we find the failure initiation mode. FMEA is a systematic tool for identifying the effects or consequences of a potential failure, methods to eliminate or reduce the chance of failure. FMEA generates a document that can be used to anticipate failure and prevent failure from occurring. FMEA can be carried out at different stages of design of a new product, in order to identify and rectify various failure modes that may be possible to make the component function as per its design intent. FMEA is generally done to check the possible failure preventions in the system rather than detecting the possible failure.

4.9.1 FMEA Methodology

- Define the system to be evaluated.
- Identify the functional relationship of the parts/components of the system and their performance requirements.
- Establish the level of analysis.
- Identify failure modes—their causes, effects, their relative importance, and their sequence.
- Identify failure detection, rectification.
- Identify design and operating provisions against such failures.
- Summarize action report.

A process is a combination of complex systems, which needs to be broken down to small individual systems such that the interdependency of the systems is minimal to carry out FMEA studies. FMEA is generally performed on individual systems to the micro level to such an extent that are predefined before the study is conducted. The actual system is a large and complex system, and carrying out FMEA on the complete system is not feasible. We need to pick up a segment of the system such that the interdependency with others is minimal and establish the limitations of this system which are under study of FMEA. Prior to the study it is important to know the full working principles of the system as well as the interaction of various components of a system and their effect on the overall system. For example, if we do not know the mechanical working principles of the system it is difficult to actually identify the interaction between the components present in that system.

The level at which an FMEA is performed is generally based on the functional structure of the system and not on the physical components of the system. Once we have identified a system either a mechanical or an electrical system, consisting of various physical components, FMEA does not analyze the physical components of the system but only undertakes a functional study of the components of the system. Therefore, it is a functional level analysis and not a physical level analysis. In a functional structure, analysis is carried out and failure modes are expressed. For example, take a physical component whose functional features are known. In case that any one of the features does not work properly as per the design intent then it is taken to be a failure mode. Failure modes are expressed as failure to perform a particular function and not as physical damage of the component. All possible failure modes should be then considered in the analysis; a few of which are listed below:

Premature operation: For example, a valve is to open when the temperature reaches $t°$, but there is a possibility that the valve may open before the temperature is reached. This is a spurious operation and referred as premature operation.

Failure to operate when required: For example, a pressure relief valve, which normally opens and releases the pressure only when the pressure inside a vessel reaches a value of "x" bar. If the pressure relief valve does not operate despite the pressure reaching "x" bar, then we can say that the valve (component) has failed to operate when it is required to operate.

Intermittent operation: For example, a specific component is supposed not required to operate at a specific temperature and pressure, however the component starts functioning physically, leading to unusual intervention to the whole process. This is considered as intermittent operation.

Failure to cease operation: For example, a temperature sensor is designed to indicate a drop of temperature below a specific value and shall remain functional until the temperature is increased. But, if it fails to stop its function on the increase of temperature, then it is a case of failure to cease operation. It should operate only when required; on other conditions, it should not remain operational.

Loss of output: For example, a sensor that is supposed to give an output, but is not available, causing loss of output during operation.

All the above are different failure modes that one needs to consider in FMEA. After identifying the various failure modes, likely causes and their effects need to be studied. The cause and effects are studied on the both the components concerned and on the overall performance of this system. We need to know what is the cause and effect of failure mode of the component, and what is the sequential effect of that failure of a component on the overall performance of the system. That is what FMEA is all about. In order to know the cause and effect of failure we start analyzing the system from the micro to the macro level. There is always a sequence of failure in the operation; consideration is given to the relative importance of the effects and the sequence in which they occur. In case there exist safeguards in the system, such a system needs to be examined separately.

TABLE 4.5
Format of Reporting FMEA Study

Component	Failure Mode	Failure Effect(s)	Comment

Amongst the identified failure modes, select the most significant failure, significant with respect to its effects on the overall system. For example, there are different failure modes. Rank them relative to one another such that which mode has a significant effect over the others on the overall performance of the system is noted. Such a mode needs to be considered and checked for whether the existing safeguards and detection devices are adequate to trace such failure in advance. As the failure mode is considered, identify what we call a weak link. A weak link is the one that has the highest rank of failure; for those weak links, one needs to carryout detailed analysis. This helps us in redesigning the weak link(s) to reduce the probability of failure of the system. In FMEA, we have a meritorious advantage of system analysis in that in a given system after identifying different failure modes one can really narrow down to what we call a weak link in the given system. Once we are able to identify the weak link, then attempt a detailed analysis on the weak link components or try to redesign those components, so that the performance level of the overall system can be increased. FMEA is reported in simple tabular form, as shown in Table 4.5.

4.9.2 FMEA Applications

FMEA is performed at various stages of development, that is, design stage and process stage. Design FMEA examines the functions of a component, the subsystem or a main system and subsequently identifies the potential failures based on improper material choice, inappropriate specification, etc. Process FMEA examines the process involved in the production line of a process industry like oil refineries. FMEA, like HaZop also has keywords describing the function of the component, aiding the study process. Basic functions are about what the product does and secondary functions are about what the process does. So, one is a design intent and other is the deviation.

In FMEA, first define the system for which one has to perform the analysis. Then identify the different failure modes in that system. Subsequently, identify the effects of the failure modes in that system where failure mode is the physical description of a failure. Identify the failure effects, which are the consequences of that failure on human safety, equipment, etc. Then their severity and the probability of occurrence of failure are identified. As it is well understood, risk is calculated at this stage of FMEA. Further identify the causes of failure and determine the occurrences of those

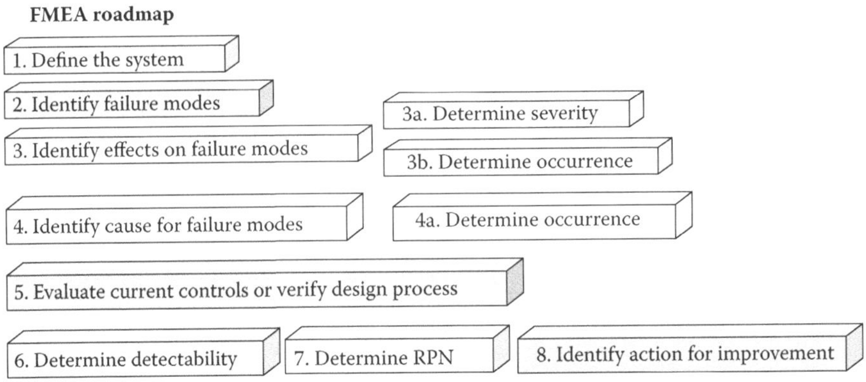

FIGURE 4.3 FMEA road map.

failure modes to evaluate the control measures that are currently available to offset the failure or redesign the process. It is important to note that FMEA is also capable of identifying any detectability present in the system design, which can improve safety or reduce risk. Subsequently, one can compute the RPN. As discussed, failures can be consequential. Therefore, effects of any specific failure or any specific failure mode on the system may not be same as that of other failure modes. Similarly, one failure mode may have a severe consequence on the system while others may not. Hence, it is necessary to prioritize the risk using the RPN. At this stage of analysis, it is clear that the qualitative features of failure mode are being converted into quantitative numbers by means of the RPN. On the basis of the RPN, one can plan for risk mitigation or risk elimination. This can result even in modifying the system design or process line, whichever results in improvement of safety. Figure 4.3 shows the FMEA road map.

4.9.3 FMEA Variables

Severity is a variable, which is a rating corresponding to the seriousness of an effect on a potential failure. It is expressed as a number on a 10-point scale. Severity of 1 indicates that it has minor effect on the system, while 10 indicates the maximum hazardous effect on the system.

Occurrence is a variable, which is a rating corresponding to the first level cause, and its resultant failure. First level cause of any specific failure will occur in the design life of the system or the product before any additional process controls are applied. Occurrence is measured on a 10-point scale. Occurrence of 1 indicates that the failure is very unlikely and 10 means the failure is certain.

Detection is a variable, which corresponds to the likelihood that the detection methods or the current controls will detect the potential failure in the design or process line before it leaves the production facility. Detection is measured on a 10-point scale; 1 refers to status that there are measures that can detect failure for certain while 10 indicates that the failure cannot be detected.

Once the above variables are quantified either based on the data of earlier failure scenarios or by experience, then one can compute the RPN, which is the product of three variables. RPN actually identifies the important areas of concern that need to be focused on the emergency front.

$$\text{RPN} = \text{Severity rating} \times \text{Occurrence rating} \times \text{Detection rating}$$

Corrective actions are taken if the severity is 9 or 10 or even if the high RPN in indicated. FMEA studies are often viewed on a relative scale; normally the highest RPN is first addressed in the analysis and design. FMEA is prepared in worksheet form, where we write down the function or the process, the possible failure modes and their effects, severity, occurrences of the failure mode, and the potential cause. Once the actions are implemented, reevaluate the new numbers for severity, occurrence, and detections, therefore obtain the new RPN to do risk assessment. Table 4.6 shows the detailed FMEA analysis, where more details are added to arrive at the RPN.

4.10 FAULT TREE AND EVENT TREE

Different sources of risk for an engineering system and activities can be analyzed with respect to their chronological order of occurrence and components that are responsible for consequences using logical trees. Logical trees are useful in analyzing the overall risk and assessment of risk contribution that arise from individual components. Fault tree and event tree diagrams are well known and the most widely applied types of logical trees, both in qualitative and quantitative risk analyses (Venkata Kiran, 2011; William, 1992). Fault trees and event trees, in many ways, are similar to each other; the choice of using either of them depends more on the traditions/preferences within a given industry than the specific characteristics of the logical tree (Young-Do and Bum, 2005). However, a significant difference between the two types of trees is that fault trees take basis in deductive (looking backwards) logic and event trees are inductive (looking forward). In practical applications, a combination of fault and event trees is typically used. In such combinations, the fault tree is used to address the sequences of failures, which may lead to events with consequences. The vent tree is used to represent the subsequent evolution of the consequence inducing events. Intersection between the fault tree and the event tree is, in reality a matter of preference of the engineer performing the study. A small event tree/large fault tree or vice versa may be applied to the same problem to supplement each other and provide additional insight with regard to the reliability of the considered system. Decision trees are often seen as a special type of event tree, but may be seen in a much wider perspective. If applied consistently within the framework of decision theory, they provide the theoretical basis for risk analysis. Detailed analyses of various types of logical trees prelude a condition that the performance of individual components of the trees have already been assessed in terms of failure rates and probabilities.

TABLE 4.6
FMEA Worksheet, Modified with Details

Part or Process Name						Supplies and Plants Affected						Person Responsible for Action Control				
Design Responsibility						Model Date						FMEA Date				
Other Areas Involved						Engg Change Level						Action Results				
Function or Process	Failure Mode	Effects	SEV (S)	OCC (O)	Potential Cause of Failure	OCC	Controls	DET	(SO)	RPN	Rec Act	Act taken	SEV	OCC	DET	RPN

4.11 FAULT TREE ANALYSIS

As mentioned earlier a fault tree is based on deductive logic. It is initiated by considering an event of system failure and then aims to deduct which causal sequence of the failure of the components could lead to the system failure. The system is thus often referred to as a top event. Logical interrelation of the sequences of component failures is represented through logical connections (logical gates). The fault tree is constructed in a tree-like structure with the top event at the top and basic events at its extremities. The basic events are those events, for which failure rate data or failure probabilities are available and which cannot be dissected further. Sometimes the events are differentiated into initiating (or triggering) events and enabling events. Initiating events are always the first event in a sequence of the enabling events while the enabling events are those which may increase the severity of the initiated failure. The fault tree is a Boolean logical diagram comprised primarily of AND and OR gates. An output event of an AND gate occurs only if all of the input events occur simultaneously and an output event of an OR gate occurs if any one of the input events occur. Figure 4.4 shows the logical gates used in the FTA.

Top events and basic events also have their specific symbols as shown in Figure 4.5.

In the figure above, the diamond-shaped symbol represents an undeveloped scenario, which is not capable of intuiting the development of sub events due to a lack of information and data. Figure 4.6 shows a typical combination of logical gates to draw a fault tree.

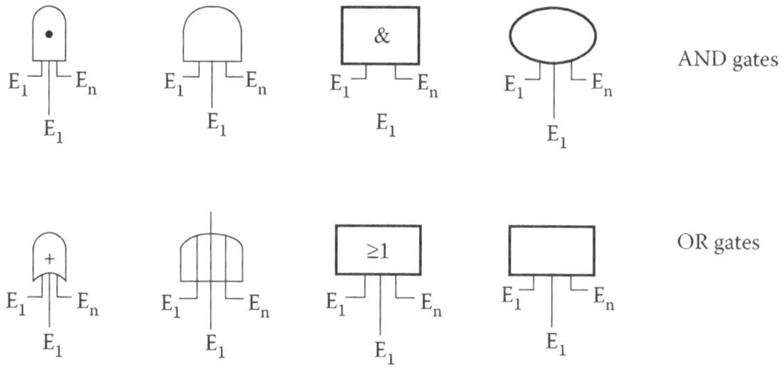

FIGURE 4.4 Logical gates for FTA.

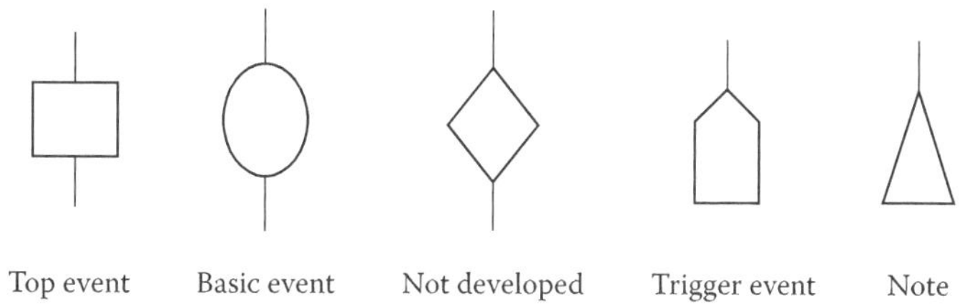

FIGURE 4.5 Symbols used in FTA.

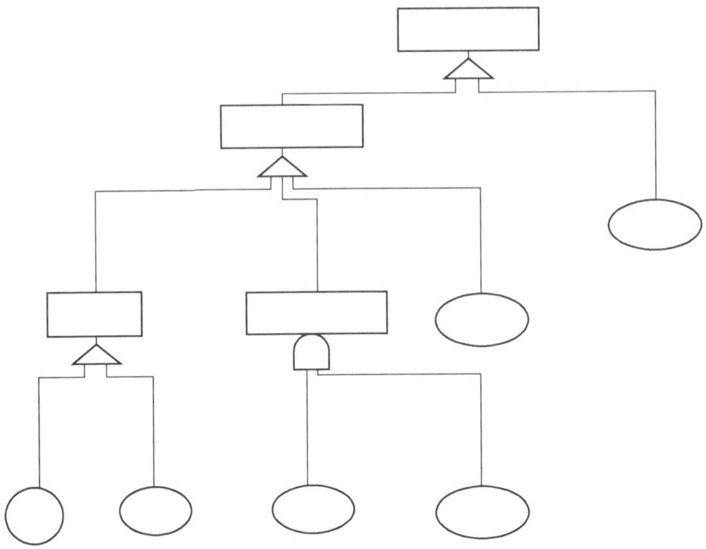

FIGURE 4.6 Typical combination of logical gates of FTA.

It is noted that a fault tree comprising an AND gate represents a parallel system, that is, all components must fail for the system to fail. Such a system thus represents some degree of redundancy because the system will still function even after one of the components fail. Fault trees comprising an OR gate, on the other hand represents a series system, that is, a system without any redundancy in the sense that it fails as soon as any one of its components fail. Such as system is often denoted a weakest component system. Systems may be represented alternatively by reliability block diagrams, as shown in Figure 4.7.

In accordance with the rules of probability theory, probability of the event for an AND gate and OR gate are given by

$$P = \prod_{i=1}^{n} p_i$$
$$P = 1 - \prod_{i=1}^{n} (1 - p_i) \tag{4.1}$$

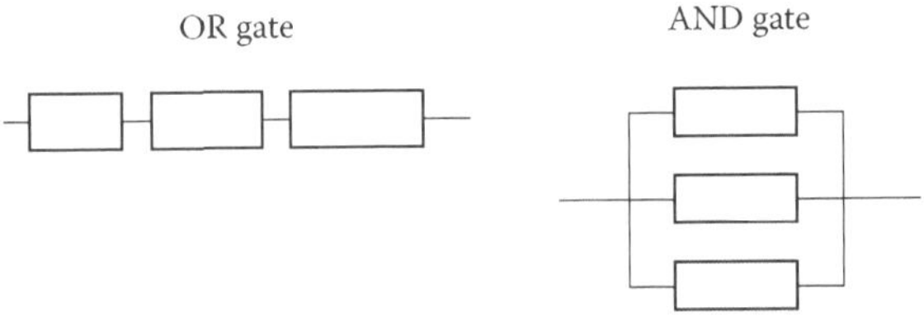

FIGURE 4.7 Block diagram for system reliability.

where n is the number of events connected to the gate, p_i are the probabilities of the failure of the events. It is assumed that the events are independent. System failure modes are defined by cut sets, which are combinations of basic events; these events, with some certainty will lead to the top event. The number of such combinations can be rather large: several hundreds for a logical tree with about 50 basic events. It is important to note that the top event may still occur even though not all basic events in a cut set occur. A minimal cut set is the cut set that represents the smallest combination of basic events leading to the top event, which is also denoted as the critical path. The top event will only occur if all the events fail in the minimal cut set. An important aspect of FTA is the identification of the minimal cut sets as this greatly facilitates the numerical evaluations involved.

4.12 EVENT TREE ANALYSIS

An event tree is a representation of the logical order of events leading to some (normally adverse) condition of interest for a considered system. It should be noted that several different states for the considered system could be associated with important consequences. In contrast to the fault tree, the event tree starts from a basic initiating event and develops from there in time until all possible states with adverse consequences have been reached. The initiating events may typically arise as top events from the FTA. The event tree is constructed from the event definitions and logical vertices (outcomes of events), which may have a discrete sample space as well as a continuous sample space. Typical graphical representations of event trees are shown in Figure 4.8. Event trees can become rather complex to analyze, which can be easily realized by noting that for a system with n two-state components, the total number of paths is 2^n. If each component has m states, then the total number of branches is m^n, which makes the analysis more complicated.

4.13 CAUSE–CONSEQUENCE ANALYSIS

Cause–consequence charts are essentially another method of representing combined fault trees and event trees. In this case, the interrelation between the fault tree and the event tree, namely the top event for the fault tree (or the initiating event for the event tree) is represented by a rectangular gate with the output event being either YES or

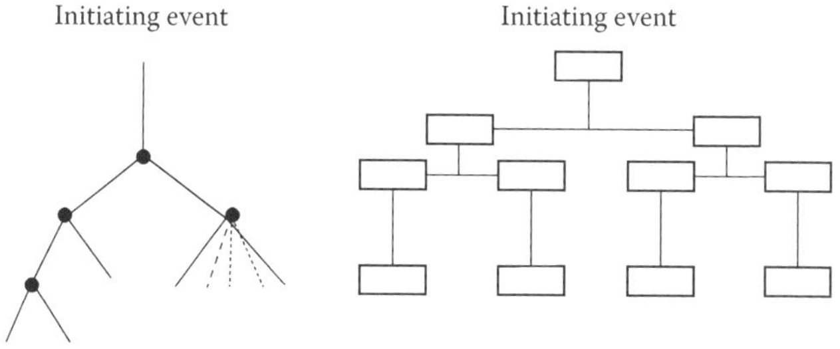

FIGURE 4.8 Event tree.

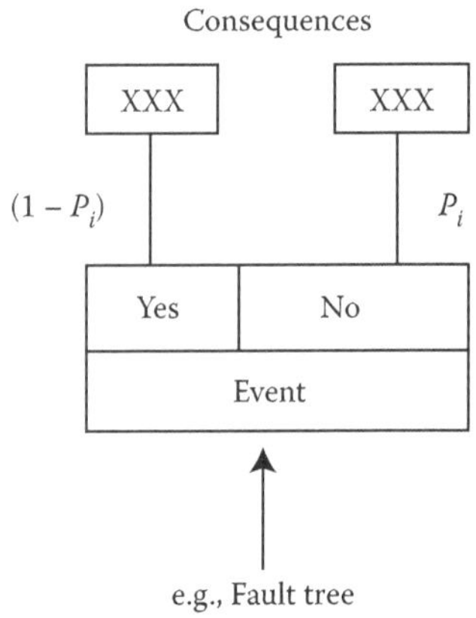

FIGURE 4.9 Cause–consequence chart.

NO. Each of these outputs will lead to a different set of consequences. The benefit of the cause–consequence chart is that the fault tree need not be expanded in the representation, enhancing the overview of the risk analysis greatly. An example of a gate in a cause–consequence chart is shown in Figure 4.9.

4.14 DECISION TREES

Decision trees are applied within the framework of decision theory, which is also one of the basic frameworks of risk assessment. This is due to the fact that risk analysis serves the purpose of decision-making and reliability which is more toward engineering judgment. Followed by the detailed risk analysis, one has to plan for risk mitigation or risk reduction, if the risk is beyond the predefined acceptable level. It is very important to note that risk level in the offshore industry should be predefined as per international practice and also needs to be declared in the public domain for approval from a competent authority. Therefore, either to act on the risk assessment results or not is a decision-making process or therefore decision trees are very much part of risk assessment. Decision analysis is the framework for risk assessment and risk evaluation as well. Figure 4.10 shows a typical decision tree used in risk assessment.

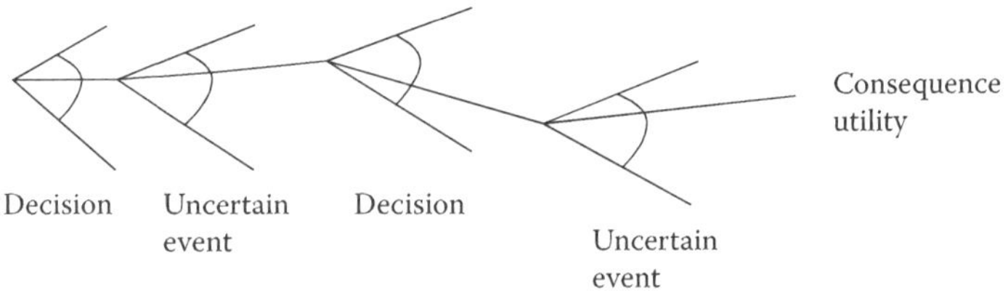

FIGURE 4.10 Decision tree for risk assessment.

A typical decision tree is constructed as a consecutive row of decisions, which is followed by uncertain events. This reflects the uncertain outcome of the possible actions that may follow from the decisions. At the end of the decision tree, consequences (or utilities) are assigned in accordance with the decisions and the outcomes of the uncertain events. Decision trees are classified depending on the number of decisions and (or) actions involved in the decision analysis. Their classification represents the various types of decision analysis that are required, ranging from the most simple ones to the most advanced. Simple ones are called prior decision analysis and the advanced are called pre-posterior analysis. It is important to note that the probabilities for different events that are represented in the decision tree may be assessed by FTA, ETA, reliability analysis, or their combination. Thus, a decision tree includes all the aspects of systems and component modeling in addition to providing a framework for decision making.

4.15 CONSEQUENCE ANALYSIS

In consequence analysis, a variety of models are used to estimate the physical effects of an accident (spill of hazardous material) and to predict the damage (lethality, injury, material destruction) of the effects. Accidental release of flammable liquids can result in severe consequences. In particular, in offshore platforms that are meant for exploration and production, even minor accidents can cause serious consequences that may affect the effective functioning of the offshore platform. For example, an immediate ignition of the pressurized chemical will result in a jet flame while a delayed ignition of flammable vapors can result in blast overpressures, which can cover a large area of spread. The consequence calculations can be roughly divided in three parts namely: (i) determination of the source-strength parameters; (ii) determination of the consequential effects; and (iii) determination of the damage or damage distances (hazard distance).

4.15.1 Source-Strength Parameters

Source-strength parameters depend on the condition in which an accident or explosion takes place. In case of the outflow of liquid vapors out of a vessel, tank, or a pipe, the volume and rate of outflow is calculated in addition to their phase of flow. In case of liquid outflow, the instantaneous flash evaporation and the dimensions of the remaining liquid pool are calculated. Further, the evaporation rate, as a function of the volatility of the material, pool dimensions, and wind velocity are determined.

4.15.2 Consequential Effects

Consequential effects depend on various parameters. Dispersion of gaseous material in the atmosphere is determined as a function of the source-strength, relative density of gas, weather conditions, and topographical situation of the surrounding area. Intensity of heat radiation (in kW/m^2) due to a fire or a BLEVE (boiling liquid expanding vapor explosion), as a function of the distance to the source can also be obtained. Energy of vapor cloud explosions (in N/m^2), as a function of distance to the

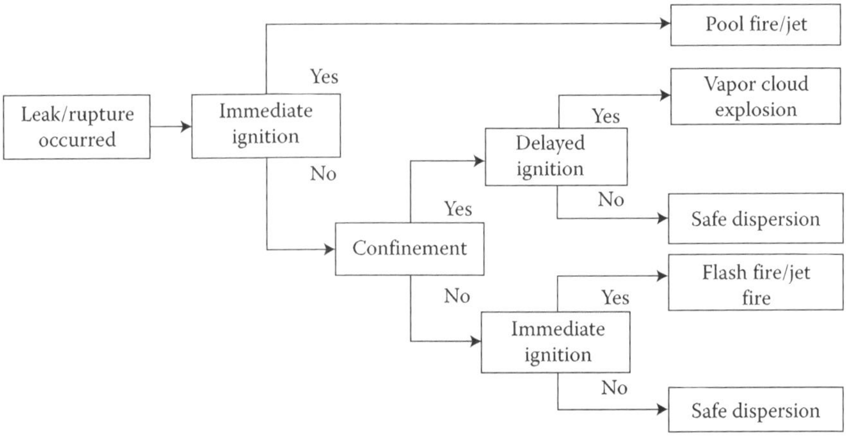

FIGURE 4.11 ETA for liquid release.

exploding cloud is computed. Concentration of gaseous material in the atmosphere, due to the dispersion of evaporated chemicals is also considered in estimating the consequences; it can be either an explosive or simply a toxic chemical release. It is obvious, that the type of models that are to be used in a risk analysis depend on the type of material involved and the way in which the material is handled. A few properties of the material and handling methods govern the choice of the risk analysis namely: phase of the material, whether they are gas, vapor, liquid, or solid; flammable characteristics of the material, whether they are explosive or toxic chemicals; storage temperature and pressure and the release type, whether they are controlled outflow (pump capacity) or catastrophic failure. Figure 4.11 shows a typical FTA of a liquid release.

4.16 LIMITATIONS OF QRA

There are a few limitations of QRA studies as different approaches give different results. The scenario selection depends on the expertise related to the specific problem. Changes in environmental conditions cause a serious deviation in the interpretation of results that arise from QRA studies. For example, operating temperature or the ambient temperature during the release of the liquid or a gas, humidity conditions present at that stage or state of release, and the wind direction and speed can alter the results which one has estimated. Each model simulates different results for the same release scenarios. If the choice of the release model is not appropriate to the above parameters, then different models project different risk assessment statements, though there is some commonness between them. In addition, the main disagreement arises from the risk acceptability criteria; acceptable risk limits vary widely for different countries. There are many countries, which even do not define acceptable risk levels for offshore industry. Therefore, the database that is used for the probability estimates can be subjective and variable as well. So, these parameters will strongly influence the results that we obtain from risk assessment studies.

4.17 RISK ACCEPTANCE CRITERIA

In the offshore industry, it is apparent that risk in unavoidable. It is not due to the fact that risk-free design and process/production is not possible. It is mainly due to the uncertainties that arise during operational conditions, mainly from the environment. Also, a typical offshore plant is one of the most complex electromechanical systems, which inherits a lot of risk by design and operation. Despite the offshore industries having a good understanding of the inbuilt risk phenomena involved in the process of oil exploration and production, it is still a difficult process to state an acceptable risk criteria. Acceptable risk does not mean risk is allowed intentionally but only to take care of any unforeseen risk that arises from the complexities of the system. Mechanical systems are very complex in design and layout. Acceptable risk is a subjective issue. A risk level which is acceptable by a regulatory agency may not be acceptable by another. A certain amount of risk is generally acceptable to regulatory agencies and also to the public. The risk acceptance level is defined by different international regulations. For example, the U.S. EPA defines risk criteria on the lifetime risk of one in a million as an acceptable one for carcinogens. According to the U.K. health and safety executive, acceptable risk is defined in terms of fatality accident rate, which recommends an acceptable value of 1.0. In general, acceptable risk is also defined (or rather should be defined) in terms of economic perspective as well. Therefore, risk is defined as "as low as reasonably practical" (ALARP), as seen in Figure 4.12.

With reference to Figure 4.13, the red band represents intolerable risk; that is risk that cannot be justified on any ground. The yellow band represents a conditional level, which is referred as the ALARP region. Risk is undertaken only if the benefit is desirable. On the other hand, the risk is tolerable only if the risk is impractical or if its cost is grossly disproportionate to the improvements that are being gained. Of course, if we have a risk in the green band then it is a broadly acceptable region and there is no need for any detailed working to demonstrate ALARP in the situation. Risk becomes completely tolerable if the cost of reduction does not exceed the improvements gained in this region. The risk criteria adopted in some countries is given in Table 4.7.

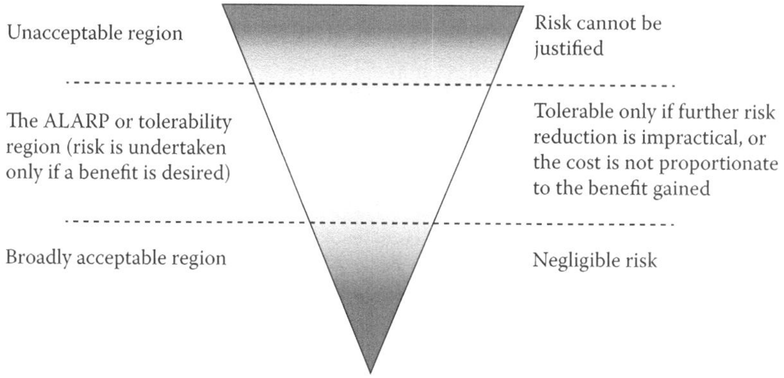

FIGURE 4.12 Risk acceptance criteria.

TABLE 4.7
Acceptable Risk Criteria

Authority and Application	Maximum Tolerable Risk (per Year)	Negligible Risk (per Year)
VROM, The Netherlands (new)	1.0E–6	1.0E–8
VROM, The Netherlands (existing)	1.0E–5	1.0E–8
HSE, UK (existing-hazardous industry)	1.0E–4	1.0E–6
HSE, UK (new nuclear power station)	1.0E–5	1.0E–6
HSE, UK (substance transport)	1.0E–4	1.0E–6
HSE, UK (new housing near plants)	1.0E–6	1.0E–7
Hong Kong Government (new plants)	1.0E–5	Not used

4.18 RISK AND HAZARD ASSESSMENT

Risk assessment is a mathematical approach practiced by engineers to predict the risk of accidents and give guidance on the appropriate means of minimizing them. It uses scientific methods and verifiable data. Hazard is a situation which can grow to become risk if not attended in advance. QRA is a very useful tool for decision-making about safety. The first stage in QRA is system definition. Defining the installation or the activity whose risks are to be analyzed is of basic importance. The scope of QRA should define the boundaries for the study, identifying which activities are included and which are excluded, and which phases of the installation's life are to be addressed. *Hazard identification* consists of a qualitative review of possible accidents that may occur, based on the experiences of previous accidents or engineering judgment wherever applicable. There are several formal techniques to perform QRA, which are useful to give a qualitative appreciation of the range and magnitude of hazards and indicate appropriate mitigation measures. In QRA, hazard identification also forms an important component, but has a more precise purpose of selecting a list of possible failure cases that are suitable for quantitative modeling. Figure 4.13 shows the flowchart of QRA. Once the hazards are identified, *frequency analysis* estimates how likely it is for the accidents to occur. Frequencies are usually obtained from analysis of previous accident experience, or by some form of theoretical modeling. In parallel with the frequency analysis, *consequence analysis* evaluates the resulting effects if the accidents occur, and their impact on personnel, equipment, and structures, the environment, or business. Estimation of the consequences of each possible event often requires some form of computer modeling, but may be based on accident experience or judgments if appropriate. When the frequencies and consequences of each modeled event are estimated, they can be combined to form measures of *overall risk*. Various forms of risk presentation may be used. Risk to life is often expressed in two complementary forms: *individual risk*, which is the risk experienced by an individual person and *societal risk*, which is the risk experienced by the whole group of people exposed to the hazard.

Subsequently, yardsticks are also identified to indicate whether the risks are acceptable, or to make some other judgment about their significance. This step begins to

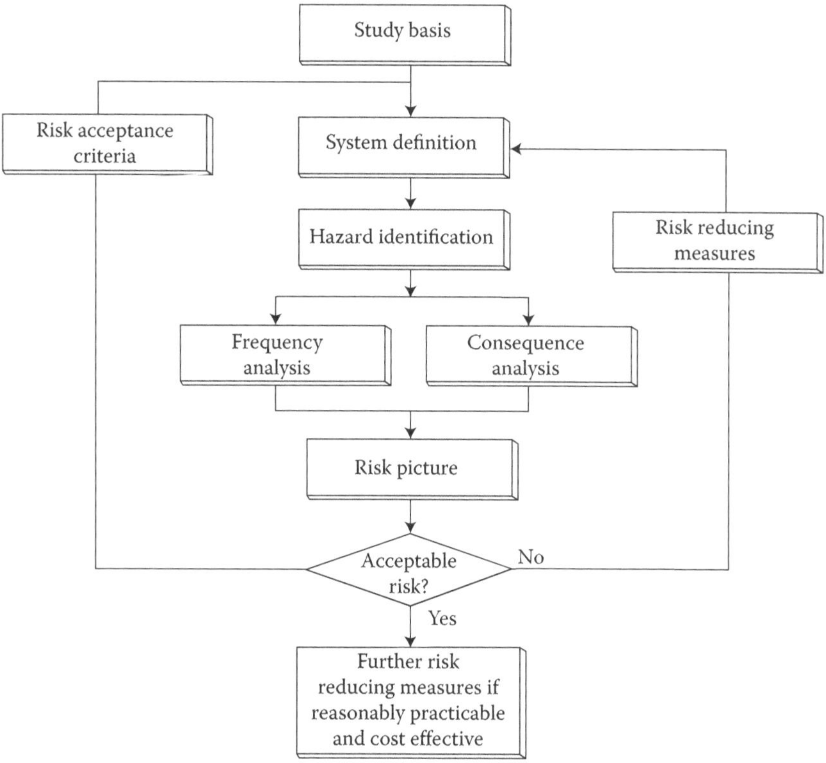

FIGURE 4.13 Flowchart of QRA.

introduce nontechnical issues of risk acceptability and decision-making, and the process is then known as *risk assessment*. In order to make the risks acceptable, *risk reduction* measures may be necessary. Benefits from these measures can be evaluated by repeating the QRA with them in place, which introduces an iterative loop into the process.

4.19 HAZARD IDENTIFICATION

In the formal risk assessment, the first step is to identify and evaluate the hazards and the unintended events, which could lead to accidents. It is important to identify the hazards that are inherent to the process and/or plant and then the evaluation of events that could be associated with hazards. The event list is generated considering various potential leaks and major releases from the rupture of all storage tanks and connected pipelines. The compiled list includes all pipe work and vessels in direct communication with the process, as they share a significant inventory which cannot be isolated in the event of emergency.

The following data are necessary to envisage various failure scenarios:

- Composition of materials stored/flowing in storage tanks/pipeline
- Inventory of materials stored in the plant

- Flow rate of materials passing through pipelines
- Operating process parameters (phase, temperature, pressure)
- Storage tank and pipeline dimensions (length, height, diameter)

The range of possible releases for a given stream covers a wide spectrum, from a pinhole leak to a catastrophic rupture (of a vessel) or full bore rupture (of a pipe). It is both time consuming and unnecessary to consider every part of the range; instead, a finite number of failure cases are generated to characterize each unit.

4.20 SELECTION OF FAILURE SCENARIOS

The selection of initiating events and incidents should consider the goals or objectives of the study and the data requirements. The data requirements increase significantly with the inclusion of non-accident initiating events and increase in the release size. While the potential range of release sizes is tremendous, groupings are both appropriate and necessitated by data restrictions. The inclusion of various release sizes and the catastrophic release is to reduce the conservative estimates in the risk analysis. They also lead to having a better understanding of the relative contributions to risk of small versus large releases. The selection of appropriate hazards considers a range of issues namely: (i) position of plant in relation to the surrounding community, (ii) complexity of the process, and (iii) nature of potential hazards. In order restrict the number of failure scenarios, the loss of containment (LOC) events, which are basically the release scenarios contributing to the individual and/or societal risk are included in the QRA. LOCs are included when one of the following two conditions is fulfilled:

- Frequency of occurrence is $\geq 10^{-9}$/year.
- Lethal damage (1% probability) occurs outside the establishment's boundary or the transport route.

There may be number of accidents that may occur quite frequently, but due to proper control measures or fewer quantities of combustible material released, they are controlled effectively. A few examples are a leak from a gasket, pump, or valve, release of a material from a vent or relief valve, and fire in a pump due to overheating. These accidents generally are controlled before they escalate by using control systems and monitoring devices that are used because such piping and equipment are known to sometimes fail or malfunction, leading to problems. After a comprehensive review of all the process facilities, failure cases in each unit are defined. Failure cases in the facilities are defined in terms of LOC scenarios; to be more precise, they address accidental releases of flammable fluids into the atmosphere. This includes various sizes of process leaks, full bore rupture, and catastrophic rupture of vessels. For each failure case, release rate and release duration are defined. This will determine the amount of material being released to the atmosphere, and hence the potential impact of the failure scenario. Duration of release is dependent on the time to detect the released fluids, time to isolate the leaking segment, and the time to discharge remaining inventory in the segment.

The total release duration is the sum of these three periods. However, time to detect depends on the following factors:

- Monitoring of process conditions, which may indicate any leak in process and/or pipeline sections
- Availability of a fire and gas detection system and/or leak detection system in a pipeline
- Surveillance of the process area, either by operator routine patrol or by a remote surveillance system
- The time to isolate is determined by the availability of an emergency shut down (ESD) system.

Various measures and systems are put in place to prevent a LOC (preventative measures) and to minimizing the effects given a LOC (mitigating measures). These are called *repression systems*. The effect of repression systems can be considered only if the effectiveness of the system is demonstrated. Common systems, which are part of repression systems are blocking systems, excess flow valves, non-return valves, bund, intervention by operators, fire protection systems, and other repression systems.

4.20.1 Blocking Systems

Blocking systems are used to limit the released quantity, following a LOC. A blocking system consists of a detection system combined with shutoff valves. The shutoff valves can be closed automatically or manually. The effectiveness of a blocking system is determined by various factors, such as position of detection monitors and their distribution throughout the various wind directions. Furthermore, the detection limit and response time of the system as well as the operator's intervention time are also relevant, which need to be considered to design a blocking system. Three different types of blocking systems are namely: (i) automatic blocking system; (ii) remote-controlled blocking system; and (iii) hand-operated blocking system. An automatic blocking system is a system where the detection of the leakage and closure of the blocking valves are fully automatic. There is no action of an operator required. Closing time of the blocking valves is about 2 minutes. In such cases, failure upon demand for the blocking system is 0.001 per demand. A remote-controlled blocking system is a system where the detection of the leakage is fully automatic. Detection of leakage results in the generation of a signal in the control room. The operator validates the signal and closes the blocking valves using a switch in the control room. Closing time of the blocking valves is about 10 minutes. Failure upon demand for the blocking system is 0.01 per demand. A hand-operated blocking system is a system where the detection of the leakage is fully automatic. The detection results in a signal in the control room. The operator validates the signal, goes to the location of the blocking valves and closes the valves by hand. Closing time of the blocking valves is about 30 minutes and the failure on demand for the blocking system is 0.01 per demand.

4.20.2 Excess Flow Valve

An excess flow valve is one that closes if the flow rate exceeds a preset value. Its operation depends on the ratio between the calculated outflow rate and the set value of the excess flow valve. Table 4.8 shows the probability of failure of the excess flow valve.

4.20.3 Non-Return Valve

A non-return valve is one that closes when the direction of the flow rate is against the preset direction. Default response time is about 5 seconds and the default probability of failure is equal to 0.06 per operation.

4.20.4 Bund

A bund is a contained or recessed area around a tank for the purpose of limiting the spread of a pool of liquid. The volume of the bund is at least equal to the storage capacity of the tank. If several tanks are present in the bund, then the volume of the bund is at least equal to the storage capacity of the largest tank plus 10% of the storage capacity of the other tanks.

4.20.5 Intervention by Operators

During a loading and unloading operation, an operator is often present on-site to supervise the process and if needed to operate shutoff valves using an emergency stop device. The intervention of an operator during loading is considered when the following conditions are met:

- The operator is present on-site from the start to the end of the loading.
- The presence of the operator on-site is guaranteed by means of a facility such as a dead man's handle or by a procedure in the safety management system and is checked during inspections.
- The process of actuating the emergency stop device by the operator present in the event of a leak during the loading operation is laid out in a procedure.
- The operator present on-site is adequately trained and is also familiar with the applicable procedures.
- The emergency stop device is positioned accordingly, so that an emergency button can be actuated within a short time irrespective of the direction of the outflow.

TABLE 4.8
Probabilities of Failure of Excess Flow Valve

Ratio of Outflow Rate and Set Value	Probability of Not Closing
Outflow rate: set value	1
Set value < outflow rate:1.2 × set value	0.12
Outflow rate > 1.2 × set value	0.06

If these conditions are met, duration of the outflow can be limited to 2 minutes in the QRA. If one of these conditions is not met, the duration of the outflow to be observed in the QRA amounts to about 30 minutes. The effective operation of an emergency stop device has a probability of failure of 0.1 per operation.

4.21 FIRE AND THERMAL RADIATION

In offshore platforms that are meant for oil exploration and production, consequences that arise from fire and thermal radiation are significantly high. In fact, fire and thermal radiation pose a major threat to offshore assets by.

4.21.1 Jet Fires

A jet fire is an intense, highly directional fire resulting from ignition of a vapor or two-phase release with significant momentum. They tend to be localized in effect and are mainly of concern in establishing the potential for domino effects and employee safety zones rather than for community risks. The jet fire model is based on the radiant fraction of total combustion energy, which is assumed to arise from a point slowly along the jet flame path. The jet dispersion model gives the jet flame length.

4.21.2 Pool Fires

A pool fire occurs when flammable liquid spillage forms a pool over a liquid or solid surface and gets ignited. Early pool fire is caused when the steady state is reached between the outflow of flammable material from the container and complete combustion of the flammable material when the ignition source is available. Late pool fires are associated with the difference between the release of material and the complete combustion of the material simultaneously. Late pool fires are common when large quantity of flammable material is released within a short time.

4.21.3 Fireball or Boiling Liquid Expanding Vapor Explosion

A BLEVE is explosively expanding vapor or two-phase fluid. A BLEVE results from a "hot rupture" of a vessel typically containing hydrocarbons such as LPG, etc., stored and maintained as a liquid under pressure, due to an impinging or engulfing fire. A flammable material will be ignited immediately upon rupture by the impinging/engulfing fire and will burn as a fireball. A fireball would also result from the immediate ignition of a release resulting from the cold catastrophic rupture of a pressurized vessel. The initial phase of a gas pipeline rupture should also be modeled as a fireball.

4.21.4 Vapor Cloud Explosion

Vapor cloud explosion (VCE) is the result of flammable materials in the atmosphere, a subsequent dispersion phase, and after some delay an ignition of the vapor cloud. Turbulence is the governing factor in blast generation, which could intensify combustion to the level that will result in an explosion. Obstacles in the path of the vapor cloud or when the cloud finds a confined area often create turbulence. The explosion results in overpressures.

4.22 SELECTION OF DAMAGE CRITERIA

The damage criteria are influenced by the relationship between the extent of physical effects (exposure) and the effect of consequences. For assessing the effects on human beings, consequences are expressed in terms of injuries and the effects on equipment/property in terms of monetary loss. In principle, two types of exposure to hazardous effects are distinguished as: (i) heat radiation due to fire (i.e., jet fires, pool fires, and BLEVE); (ii) explosion, that is, VCE.

4.22.1 Heat Radiation

The effect of fire on a human being is in the form of burns. There are three categories of burn such as first-degree, second-degree, and third-degree burns. The consequences caused by exposure to heat radiation are a function of three factors namely: (i) radiation energy onto the human body (kW/m^2); (ii) exposure duration (s); and (iii) protection of the skin tissue (clothed or naked body). Limits for 1% of the exposed people to be killed due to heat radiation, and for second-degree burns are given in Table 4.9. To estimate the consequence of heat radiation, as a part of risk assessment, a few assumptions are made namely: (i) lethality of a jet fire or a pool fire is assumed to be 100% for the people who are caught in the flame. Outside the flame area, lethality depends on the heat radiation distances; and (ii) flash fire lethality is taken as 100% for all the people caught outdoors and for 10% who are indoors within the flammable cloud. No fatality is assumed outside the flash fire area. Table 4.10 shows the damage to human life with respect to exposure to fire.

TABLE 4.9
Effect of Heat Radiation on Humans and Equipment

Heat Radiation (kW/m^2)	Damage Level	
	People	Equipment
1.6	No discomfort for long exposure	
4.0	Sufficient to cause pain within 20 s. Blistering of skin (first-degree burns are likely)	
4.7	Accepted value to represent injury	
10.0	Pain threshold reached after 8 seconds and second-degree burn after 25 seconds	
12.5	100% fatality after short time exposure	Minimum energy required for melting of plastic
25		Minimum energy required to ignite wood
37.5		Sufficient to cause major damage to equipment

TABLE 4.10
Damage to Human Life with Respect to Time of Exposure

Exposure Duration (s)	Radiation (1% Lethality) (kW/m^2)	Radiation for Second-Degree Burns (kW/m^2)	Radiation for First—Degree Burns (kW/m^2)
10	21.2	16	12.5
30	9.3	7.0	4.0

4.22.2 Explosion

In case of VCE, two physical effects may occur namely: (i) a flash fire over the whole length of the explosive gas cloud; and (ii) a blast wave, with typical peak overpressures circular around the ignition source. The effect of overpressure on human beings is twofold namely: (i) direct effect of overpressure on human organs; and (ii) effect of debris from structure damage affecting humans. The direct effect of overpressure on humans is more significant. When the pressure change is sudden, the pressure difference that arises can lead to the damage of some organs. The extent of damage varies with the overpressure along with factors such as position of the person, protection inside a shelter, body weight as well as duration of the overpressure. The organs prone to get affected by overpressure are the eardrum and lungs. Another important aspect is the effect of debris from structure damage affecting humans. Overpressure duration is an important factor for determining the effects on structures. The positive pressure phase can last for about 10–250 ms. The same overpressure can have markedly different effects depending on its duration.

For the blast wave, the lethality criterion is based on a peak overpressure of 0.1 bar that will cause serious damage to 10% of the housing/structures. It is interesting to note that falling fragments will kill one of each eight persons in the destroyed buildings. The damage criteria given in Table 4.11 indicate the effect of overpressure. While making the estimate of the effects of overpressure, a few assumptions are made: (i) overpressure more than 0.3 bar corresponds to approximately 50% lethality; (ii) overpressure above 0.2 bar would result in 10% fatalities; (iii) overpressure <0.1 bar would not cause any fatalities to the public; and (iv) 100% lethality is assumed for all people who are present within the cloud.

TABLE 4.11
Effect of Overpressure

Peak Overpressure (bar)	Damage Description
1.70	Bursting of lung
0.30	Major damage to plant equipment structure
0.20	Minor damage to steel frames
0.10	Repairable damage to plant equipment and structure
0.03	Shattering of glass
0.01	Crack in glass

4.23 RISK PICTURE

Risk is defined as a measure of potential, economic loss, or human injury in terms of the probability of the loss or injury, occurring to the public and the magnitude of the loss or injury if it occurs. Risk assessment depends on many factors, which characterize the quality of risk. They are namely: (i) population of the place/plant under consideration; and (ii) ignition probabilities. It is necessary to know the population exposure in order to estimate the consequences and the risk resulting from an incident. The exposed population is often defined using a population density. Population densities are an important part of risk assessment for several reasons. The most notable is that the density is typically used to determine the number of people affected by a given incident with a specific hazard area. Sometimes, population data are available in sketchy forms. Population density can be averaged over the whole area that may be affected. Alternatively, the area can be subdivided into any number of segments with a separate population density for each individual segment. For material releases from the storage/handling system, where a large percentage of rupture events may be due to third party damage, a relatively high probability of immediate ignition is generally used. Delayed ignition takes other factors into account. Delayed ignition probabilities can also be determined as a function of the cloud area or the location. In general as the size of the cloud increases, the probability of delayed ignition decreases.

4.24 INDIVIDUAL RISK

Risk assessment is carried out by combining the consequence of every possible event. From each failure case and human impact criteria as defined previously, event frequencies for a particular event are determined. The term "individual risk" is used for calculations of the risk of fatality for someone at a specific location. During this estimate, it is assumed that the person is always present at the location and he is continuously exposed to the risk at that location. This is sometimes referred to as location-specific individual risk (LSIR). This is explicit to distinguish it from the "person-specific individual risk" that would depend on the movements of a given individual. It is a measure of the geographic distribution of risk and is independent of the distribution of people at that location or in the surrounding area.

The individual risk $IR_{M,x,y|w}$ for a location x, y, for weather condition w is given by

$$IR_{M,x,y|w} = F_M \int_{\theta_1}^{\theta_2} [P_{\theta|w} P_{d|\theta w}\, d\theta \tag{4.2}$$

where location x, y is the center of a given calculation location, weather w a given combination of wind speed and atmospheric stability, F_M the event frequency for the scenario, θ the direction of the release, θ_1 that impacts the lower value of location x, y, θ_2 the upper value of that impacts the location x, y, $P_{\theta|w}$ the probability of the release occurring in that direction given the weather, and $P_{d|\theta w}$ is the probability of death given that release direction and weather. This is the contribution to the individual risk at that location for a given weather condition. Total contribution $IR_{M,x,y}$ is the sum of the contributions for all weather cases and is given by

$$IR_{M,x,y} = \sum_{\text{All weathers}} P_w IR_{M,x,y|w} \tag{4.3}$$

Risk results are presented in the form of risk contour plot, which shows the distribution of LSIR against the background of a map. In addition, individual risk per annum (IRPA) is calculated for each worker group. IRPA reflects the risk of an individual spending certain fractions of his/her time at specific locations at the site. The fraction of time where a particular individual spends at specific location is combined with the LSIR at the particular location to derive the IRPA, which is defined as

$$IRPA = \sum_{\text{All locations}} LSIR \cdot f_L \tag{4.4}$$

where f_L indicates the fraction of time during a year that a particular person in this group spends in the particular location. The IRPA for each group of people is calculated separately.

4.25 SOCIETAL RISK

This is a measure of risk that the events pose to the local population, taking into account the distribution of the population in the local area. Societal risk is expressed in terms of the likelihood of event outcomes that affect a given number of people in a single incident (e.g., the likelihood of event outcomes that affect up to 10 people, or the likelihood of event outcomes that affect up to 20 people). Number of fatalities $N_{M|O}$ caused for a given combination of weather condition, direction, and event outcome, is given by

$$N_{M|O} = \iint n_{x,y} P_{d,x,y|O} dxdy \tag{4.5}$$

where $n_{x,y}$ is the population density in the location whose center is at x, y, and P_{d}, x, $y|o$ is the probability of death from the effect zone produced by the outcome. Combination of the frequency $F_{M|O}$ for the location and outcome, and the number of associated fatalities $N_{M|O}$ is known as an "F–N pair," and the frequencies for given values of N can be summed for all outcomes to give the total societal risk, presented in the form of a table or curve. The rate of death is also commonly known as the potential loss of life (PLL), which is the long-term average number of fatalities per year for a group of people. The PLL is a measure of the risk to a group of people as a whole and is particularly effective in measuring the effectiveness of various risk reducing measures. The risk results are presented in the form of *FN* Curve, which shows the frequency (F) of outcomes which cause N or more fatalities. In addition, the PLL and contribution of each failure case to the societal risk are calculated.

4.26 RISK ASSESSMENT AND MANAGEMENT

Risk assessment and management is a process of *identifying* all loss exposures, *evaluating* the risk, *developing* a plan, *implementing* the plan, and *monitoring* the plan depicted pictorially as in Figure 4.14.

4.26.1 Objectives of Risk Management

Risk management is essential for establishing control on risk reduction. There are two types of objectives namely: pre-loss and post-loss objectives. Pre-loss objectives consist of the following namely: (i) economy to minimize the expenditures consistent with post-loss goals; (ii) reduction in anxiety to reduce fear and worry over potential loss; (iii) meeting externally imposed obligations leading to satisfying HSE (Health and Safety Executive) regulations, acquiring required insurances, satisfying employee-benefit plans; and (iv) social responsibility to meet the demand of good citizenship to employees, maintaining good public image, etc. Post-loss objectives include: (i) survival, which refers to the capability to resume some operation even after a loss; (ii) continuity of operations, leading to resuming full operations following an interruption. There may be reduction in earnings, keeping human and material resources available for resumption; (iii) earnings stability by earning through continued operations with cost control; (iv) continued growth by finding ways to expand growth; and (v) social responsibility toward employees care, public image, etc. Figure 4.15 shows the risk assessment matrix.

With reference to the figure above, hazard severity levels are indicated as level I–IV. (i) Level I indicates death or system loss, (ii) level II indicates severe injury or major system damage; (iii) level III indicates minor injury or minor system

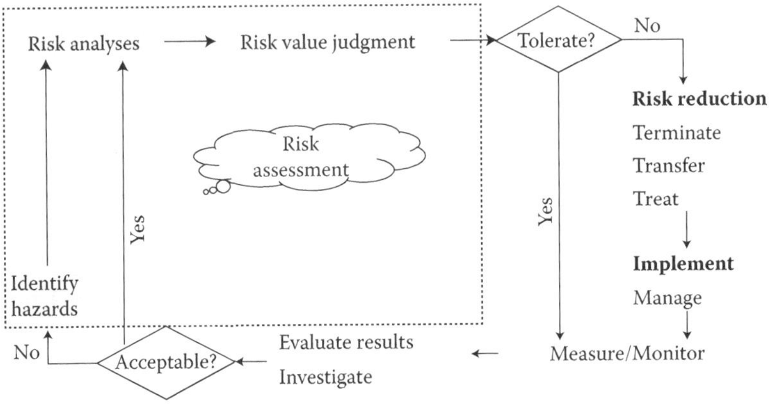

FIGURE 4.14 Risk assessment and management.

Hazard severity	Probability				
	Frequent	Probable	Occasional	Remote	Improbable
Catastrophic (I)	Risk reduction required				
Critical (II)	Risk reduction required				
Marginal (III)	Written waiver mgmt. required		Operation permissible		
Negligible (IV)	Operation permissible				

FIGURE 4.15 Risk assessment matrix.

damage; and (iv) level IV indicates less than minor injury or system damage. Probability levels indicate frequent as likely to occur frequently, probable as that which will occur several times in the life of an event, occasional as that which is likely to occur sometimes in life of an event, remote to be unlikely but possible to occur in the life of an event, and improbable to refer to that which is unlikely to occur in the life of an event.

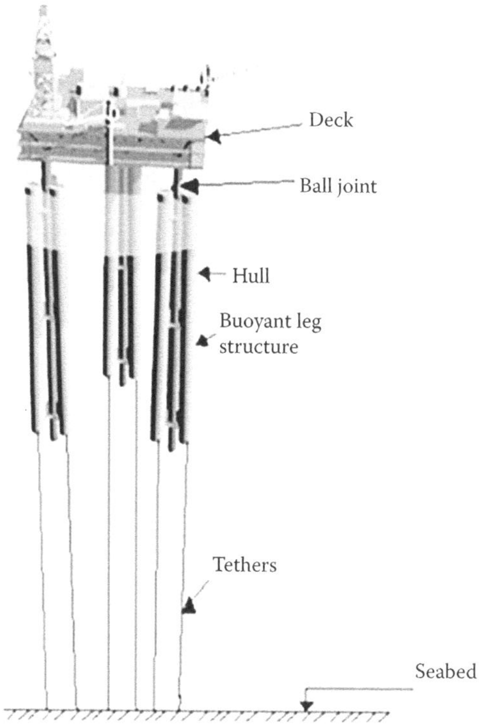

FIGURE 4.16 Offshore triceratops.

4.27 EXAMPLE PROBLEM OF RISK ASSESSMENT: OFFSHORE TRICERATOPS

As oil exploration is heading toward deep and ultra-deepwater depths, more innovative and simple structures are essential. The concept of triceratops is new for dry tree-based developments. Offshore triceratops consist of topside, three buoyant leg structures (BLS), ball joint between deck and BLS, and foundation system with tethers. The main advantage of the ball joint is that it transfers translations but not rotations from the substructure to the super structure and vice versa. The offshore triceratops have a compliant, deep-draft, positively buoyant structure, which is secured to the seafloor either with restraining system or tethers. It is simple in

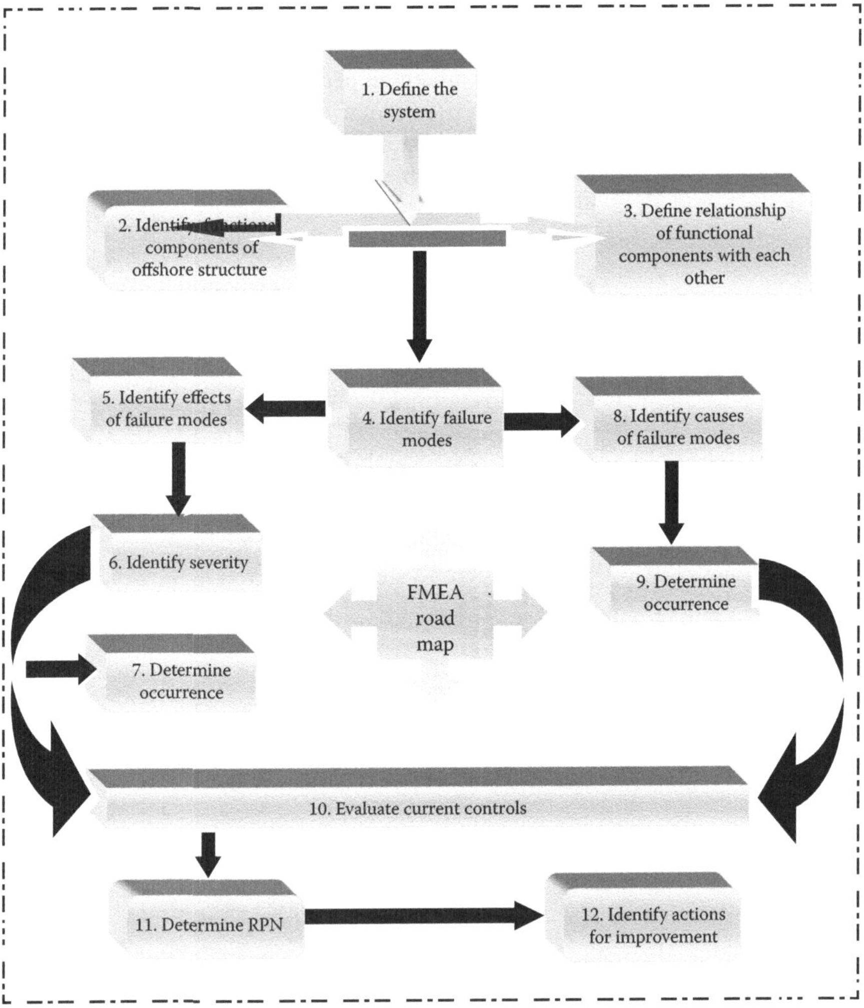

FIGURE 4.17 FMEA methodology used for example problem.

geometric form and gains required buoyancy due to its underwater volume. It supports the deck loads, structure mass, ballast, facilities, and pretension. Figure 4.16 shows offshore triceratops used for deepwater oil exploration.

Triceratops, supported by BLS appears to be a spar platform but behaves like a tension leg platform due to its restraining system when tethers are used. The concept of BLS was introduced (Copple and Capanoglu, 1995; Shaver et al., 2001), and preliminary analysis and experiments were performed. The analytical and experimental results were validated with some discrepancies. Some experimental and analytical comparative studies (Capanoglu et al., 2002) are performed for BLS with restraining leg; the results are with some discrepancies. Further, the BLS is upgraded to triceratops. The experimental investigation has been performed (Srinivasan Chandrasekaran and Madhuri, 2015) and the advantage of the ball joint was presented in pitch response amplitude operators (RAO). Preliminary experimental investigations that were carried out on the scaled model of triceratops show advantageous features that suit deepwater conditions (Srinivasan Chandrasekaran et al., 2011a). Recent studies carried out show advantageous adaptability of offshore

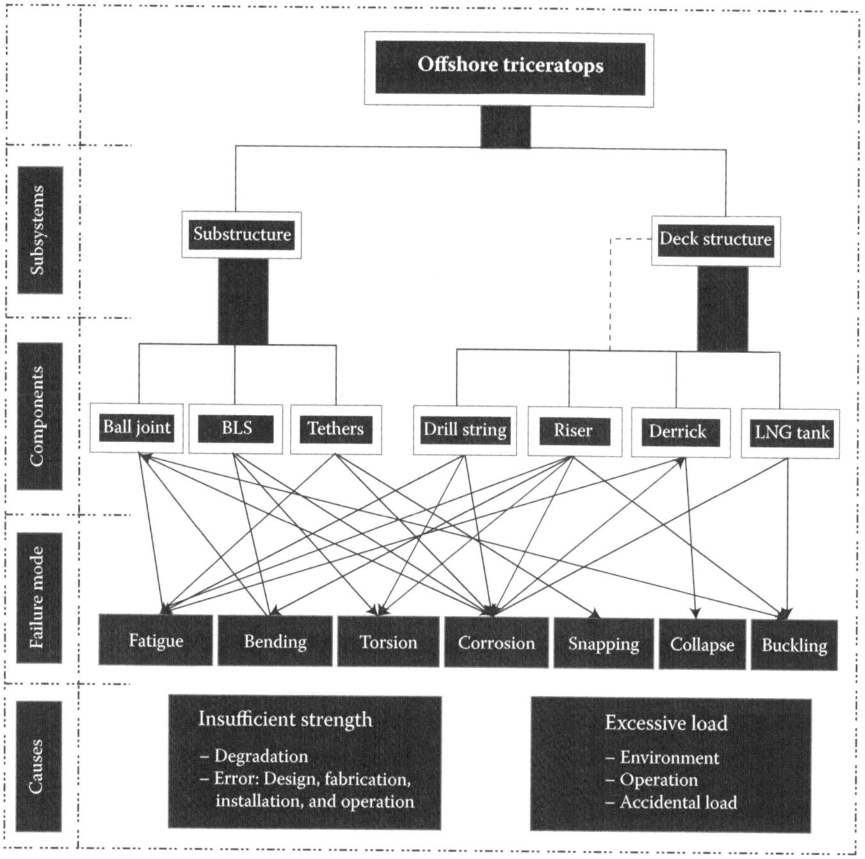

FIGURE 4.18 Component breakdown and failure modes of offshore triceratops.

TABLE 4.12
Risk Analysis Summary for Offshore Triceratops

S/N	Components	Failure Mode	Failure Effect	Causes/Sources	Recommended Actions
1	Ball joint	Fatigue, corrosion, bending, buckling	Cracks, misalignment, collapse of the entire structure, fatality	Faulty design, manufacturing defect, biological, environmental factors, improper lubrication	• Rigorous testing required • Proper lubrication required • Proper material selection
2	BLS	Bending, torsion, corrosion	Instability, overturning, submerging, fatality	Environment factors, design, and installation error	• Proper inspection required • Ultrasonic welding required to avoid welding flaws
3	Tethers (umbilical)	Fatigue, corrosion, snapping	Instability	Manufacturing defects, environmental factors, accident	• Installation of sensors to send signals to approaching objects • Proper material selection • Proper manufacturing skills
4	Drill string	Fatigue, torsion, corrosion	Oil spillage, consume more energy	Environmental factors, operation	• Adherence to safety standards • Routine checks required for failure detection • Proper material selection method
5	Risers	Fatigue, corrosion, bending, buckling	Oil spillage, affects operation	Environmental factors, operation effect	• Proper material selection method • Adherence to safety standards • Routine checks to detect failure initiation
6	Derrick	Fatigue, corrosion, collapse	Closure of entire operation, fatality	Environmental factors, poor material selection, faulty design	• Adherence to manufacturing standards • Proper inspection to detect failure initiation • Anticorrosion coating required
7	LNG tank	Buckling, corrosion	Leakage, explosion, fatality	Poor material, selection	• Proper material selection required • Proper coating against corrosion

TABLE 4.13
FMEA-Offshore Triceratops

Prepared by: XXX **Design Responsibility: XXX company**

FMEA No: 01 **Date: 20/12/2014**

						Failure Causes						
Component	**Function**	**Failure Mode (s)**	**Effects**	**SEV (S)**	**OCC (O)**	**Potential Reasons**	**OCC**	**Controls**	**Dt**	**S×O**	**RPN**	**Recommended Actions**
Ball joint(s)	Support deck weight, connects deck to BLS	Fatigue Collision Binding Bucking	Cracks, misalignment, collapse of the entire structure, fatality	4	4	Faulty design, manufacturing defect, biological, environmental factors. improper lubrication	3	Routine checks and adoption of proper standards	7	16	112	• Rigorous testing required • Proper lubrication required • Proper material selection
BLS	Provides buoyancy	Bending Torsion Corrosion	Instability, overturning submerging, fatality	4	3	Environment factors, design and installation error	3	Check BLS design and buoyancy properties	6	12	72	• Proper inspection required • Ultrasonic welding required to avoid welding flaws
Tether(s)	Provides tension	Fatigue Corrosion Snapping	Instability	3	3	Manufacturing defects, environmental factors, accident	3	Check weaving uniformity and material properties	7	9	63	• Installation of sensors to send signals to approaching objects • Proper material selection • Proper manufacturing skills
Drill string	Suction of hydrocarbon	Fatigue Torsion Corrosion	Oil spillage, consumer more energy	5	2	Environmental factors, operation	6	Proper lubrication and proper checks	7	10	70	• Adherence to safety standards • Routine checks required for failure detection

(Continued)

TABLE 4.13 (*Continued*)
FMEA-Offshore Triceratops

Prepared by: XXX | **Design Responsibility: XXX company**

FMEA No: 01 | **Date: 20/12/2014**

Component	Function	Failure Mode (s)	Effects	SEV (S)	OCC (O)	Failure Causes: Potential Reasons	Failure Causes: OCC	Controls	Dt	S×O	RPN	Recommended Actions
Risers	Houses drill holds drill bit	Fatigue Corrosion Bending Buckling	Oil spillage, affects operation	5	2	Environmental factors, operation effect	6	Routine checks and maintenance before each operation	6	10	60	• Proper material selection method • Adherence to safety standard • Routine checks to detect failure initiation
Derrick	Holds riser during drilling	Fatigue Corrosion Collapse	Closure of entire operation, fatality	4	2	Environmental factors, poor material selection, faulty design	2	Routine checks and maintenance	6	8	48	• Adherence to manufacturing standards • Proper inspection to detect failure initiation • Anticorrosion coating required
LNG tank	Stores hydrocarbon	Buckling Corrosion	Leakage, explosion	4	2	Poor material selection	2	Proper gauge installation	7	8	56	• Proper material selection • Proper coating against corrosion

triceratops for ultra-deepwater oil exploration. As the structural form is relatively new, it is necessary to do a risk assessment for the structural form. In the present example, it is proposed to carry out a risk assessment using the FMEA tool. Figure 4.17 shows the FMEA methodology used in the current example.

Prior to the risk assessment, it is necessary to perform a detailed analysis to understand the probabilities of component level breakdown and various failure modes that may arise during the commissioning and operation of the new platform. Figure 4.18 shows the layout of component level breakdown and failure modes, identified for offshore triceratops.

Risk analysis is carried out to supply vital information for the FMEA study. The results of the risk analysis are shown in Table 4.12. As explained earlier, a detailed FMEA is carried out and the results of the risk assessment are shown in Table 4.13.

Model Exercise Papers

EXERCISE PAPER 1

PART A

Answer all questions and to the point. Every step carries credits.

Question 1

a. Define:
   i. "Measure of Belief" probability and its drawbacks.
   ii. Suppose a statement is made that: "The probability of a football match to be won by Germany is 0.9 for a match with India." Does this statement supersede the classical definition? If so, how.

b. Suppose the data below gives the ordered first fatigue load and final failure load for a K-joint of a jacket. Calculate the mean and standard deviation of the data. Which one has more sample coefficient of variation? (Hint: Use sample variance and not population variance.)

First-Failure Load (kN)	Final-Failure Load (kN)
5100	9300
5800	9300
6000	9400
6000	9500
5800	9500

c. What are the SI units of the following
   (a) Mean, (b) variance, (c) standard deviation, and (d) kurtosis.

d. What is a random variable?

e. Name one cumulative distribution that describes: discrete random variable and continuous random variable. State one process each which best describes these random variables.

f. What do you mean by independence of a random variable?

Question 2

Bayes' theorem: An existing offshore structure is to be tested for reuse so that the decommission process is not required to be done. Therefore, the structure was analyzed using the numerical software ANSYS-AQWA. Prior to starting the analysis, the designer decides that the structure would be marked against three types of failure (Able to be Reused, Minor Repair Necessary, and To be Changed) based on his judgment and knowledge of similar structures.

State of Failure	Event	Probability
Able to Be Reused	A	0.2
Minor Repair Necessary	B	0.7
To Be Changed	C	0.1

The above table means that if the structure is analyzed 70% of the time the structure can be used after minor repairs. Thirty percent of the time the parts of the structure can either be reused or requires change.

Since the FEM calculations using ANSYS-AQWA also have some limitations as to the order of calculation as well as the numerical methods used. So he needs to assign additional conditional probabilities to the analysis method by also using the software. After the ANSYS-AQWA analysis, the states were again predetermined and the following results were obtained:

	Event		
ANSYS-AQWA ↓	A	B	C
$\bar{A}$ (Able to Be Reused)	$P(\bar{A} \mid A) = 0.7$	$P(\bar{A} \mid B) = 0.2$	$P(\bar{A} \mid C) = 0$
$\bar{B}$ (Minor Repair Necessary)	$P(\bar{B} \mid A) = 0.3$	$P(\bar{B} \mid B) = 0.6$	$P(\bar{B} \mid C) = 0.3$
$\bar{C}$ (To Be Changed)	$P(\bar{C} \mid A) = 0.0$	$P(\bar{C} \mid B) = 0.2$	$P(\bar{C} \mid C) = 0.7$
	Sum = 1.0	Sum = 1.0	Sum = 1.0

The above table indicates that if one is looking for the structure to be "Able to be Reused," then after ANSYS-AQWA calculations one can be sure that 70% of the time the analysis yields "Able to be Reused," 20% of time the structure might be in the event "Minor Repair Necessary."

Calculate using Total Probability Theorem $P(A \mid \bar{A})$, $P(B \mid \bar{A})$, $P(B \mid \bar{B})$

Hint: The Total Probability Theorem is given by

$$P(A \mid \bar{B}) = \frac{P(\bar{B} \mid A)P(A)}{P(\bar{B} \mid A)P(A) + P(\bar{B} \mid B)P(B) + P(\bar{B} \mid C)P(C)}.$$

Part B

Answer all questions

1. Define risk and reliability, highlighting their differences.
2. State different levels of reliability and briefly explain their suitability for different kinds of engineering problems.
3. What are uncertainties encountered in probabilistic analysis of offshore structures? How are they accounted for in the analysis?
4. In the test conducted on concrete mix design, it is concluded that 30% of coarse aggregate (CA), 25% of fine aggregate (FA), 25% of cement, and 20% of water generally control the overall strength. These data were arrived based on experiments conducted in the lab. However, it is understood that

not all the samples tested were reliable and therefore one cannot rely on the statement made for any further assessment. It is now necessary to examine the adequacy of the test results; a sample having the following data is to be assessed. CA = 27%; FA = 10%; Cement = 30%; water = 50%. Estimate the reliability of information from the sampling on one of the random samples?

5. Hundred concrete cubes are tested for the compressive strength and the test data are as given below: Draw the pdf and cdf for the given data:

f_{ck} (N/mm²)	Frequency of the Class
24–25	7
25–26	15
26–27	20
27–28	25
28–29	15
29–30	10
30–31	8

EXERCISE PAPER 2

Part A

Answer all questions and to the point. Write each step clearly.

Question 1

a. You are given the contours of the joint density function $f_{XY}(x, y)$. Given a value y_0, show in the following figure $f_{XY}(x|y_0)$.

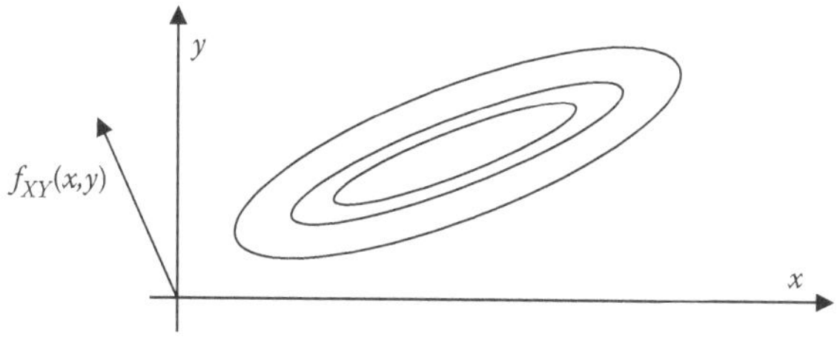

b. Use the following figures. Assume that the pdf of the random variable X is known (e.g., Gaussian) and upon nonlinear transformation one gets another random variable $Y = g(X)$. Draw the pdf for Y.

c. Write down and provide reasons for the type of $\rho_{XY} = \text{Cov}(X,Y)/\sigma_X\sigma_Y$ exhibited by the joint random variables.

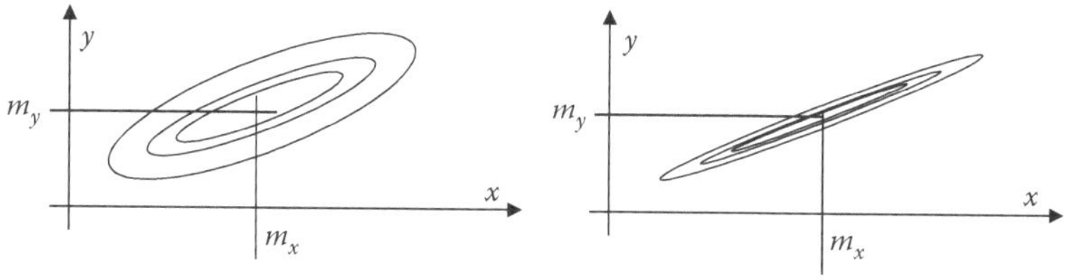

d. Explain the moments of random variables? If two variables are independent then what would be the value of μ_{XY}?

Question 2

Define the following:

a.
   i. Estimator
   ii. Estimate
   iii. Confidence interval of the mean
   iv. Probability distribution within the confidence interval of data obtained from multiple tests

b. An offshore engineer measures the impact forces on the pontoon with the assumption of Gaussian type. The Gaussian density function is given by $f_X(x;\mu,\sigma) = (1/2\pi)\exp(-(x-\mu)^2/2\sigma^2)$

Test	Load (MN)
Ship 1	450
Ship 2	475
Ship 3	800
Ship 4	781
Ship 5	785

With respect to the above information and using method of moments write the (i) The estimator, (ii) the estimate, (iii) 95% confidence interval, (iv) skewness, and (v) kurtosis.

Part B

Answer all questions

1. An offshore accident on board is reported to have occurred due to mechanical fault during shift operation. Three operators A, B, and C were on duty according to duty login. Operator A commits an error of 2% while B and C commit errors of 5% and 8%, respectively, which is seen from their past service records. Operator A was in the job for 60% of the duty time while B and C were 15% and 25%, respectively. It is concluded that accident occurred only due to the fault of the operator. What is the probability that it may be due to operator A?
2. Probability that a structure fails after service life of 20 years is 0.05. Find the probability that out of five such structures inspected, (i) none; (ii) not more than one; (iii) more than one; (iv) at least one will fail after 20 years of service life.
3. In the reliability estimate of offshore structures, it is seen that 60% of the service life is influenced by overload on the structure and 40% by material degradation, if these factors are considered independently. However, on their joint contribution, their influence on the service life is about 20%.

Given the above information as authentic, an offshore structure is selected at random for service life estimate. Find the probability that

a. The service life of the structure shall not be influenced by either of these factors
b. If the service life is influenced by overload, then find the probability that it is also influenced by material degradation
c. If the service life is influenced by material degradation, then find the probability that it is also influenced by overload

4. An insurance company insured 20 drill rig operators, 40 electricians, and 60 floor shop mechanics. The probability of accidents is 0.15, 0.03, and 0.01, respectively. One of the insured persons meets with an accident. What is the probability that he is a drill rig operator?
5. A random variable X has the following probability distribution:

X	0	1	2	3	4	5	6	7
$P(X)$	0	k	$2k$	$2k$	$3k$	K^2	$2k^2$	$7k^2 + k$

Determine

i. k
ii. $P(X < 3)$
iii. $P(X > 6)$
iv. $P(0 < X < 3)$

Part C

Answer all questions

6. Probability of failure of a structure is given by the following equation:

$$P_f = \int_{0<t\le T} P[\min g(x, y(t)) \le 0 | x] f(x)\, dx$$

Explain the various terms involved in estimating the P_f. Why exact solution of the above equation is difficult?

7. What do you understand by FOSM? Derive the equation for probability of failure using FOSM.
8. Write a brief note on the Hasofer–Lind method.
9. Write a brief note on the simulation-based reliability method.

EXERCISE PAPER 3

Answer all questions

1. A consultant inspects an offshore platform on request to estimate the service life on an "as is where is" basis. He identifies basic parameters that influence the strength on the basis of a site investigation and past experience. His assessment is that the factors are (a) overloading; (b) material strength

degradation due to age; (c) impact loads due to barges, etc. Probabilities associated with these factors are 4/10, 5/10, and 1/10, respectively. The probabilities that the accuracy of his estimate on the dependence of factors are 1/4, 1/3, and 1/2, respectively. What is the probability that the service life will depend mainly on the material degradation?

2. Space of events of a probabilistic assessment of failure is 15; each of the events is given weighted importance based on experience as 14, 17, 15, 14, 21, 17, 19, 20, 16, 18, 20, 17, 16, 19, 20. One event is selected in such a manner that each has the same chance of being chosen. Weightage X of the chosen event is recorded. What is the probability distribution of the random variable X in this study? Find the mean, variance, and standard deviation of the variable.
3. Probability that a structure fails after service life of 20 years is 0.05. Find the probability that out of five such structures inspected, (i) none; (ii) not more than one; (iii) more than one; (iv) at least one will fail after 20 years of service life.
4. An offshore accident on board is reported to have occurred due to mechanical fault during shift operation. Three operators A, B, and C were on duty via duty login. Operator A commits error of 1% while B and C commits error of 5% and 7%, respectively, which is seen from their past service records. A was in the job for 50% of the duty time while B and C were 30% and 20%, respectively. It is concluded that accident occurred only due to the fault of the operator. What is the probability that it may be due to operator A?

EXERCISE PAPER 4

Answer all questions

1. Two consultants namely A and B were shortlisted to inspect an offshore facility for assessing the service life. The required experience sought by the company is 20 years for such service life assessment, while the past experience of A and B are only 10 and 6.5 years, respectively. If both the consultants try to assess the offshore facility independently, find the probability that (i) the service life shall be estimated correctly; (ii) exactly one of the consultants shall estimate it correctly.

   In the reliability estimate of offshore structures, it is seen that 60% of the service life is influenced by overload on the structure and 40% by material degradation, if these factors are considered independently. However, on their joint contribution, their influence on the service life is about 20%. Given the above information as authentic, an offshore structure is selected at random for service life estimate. Find the probability that

   a. The service life of the structure shall not be influenced by either of these factors
   b. If the service life is influenced by overload, then find the probability that it is also influenced by material degradation
   c. If the service life is influenced by material degradation, then find the probability that it is also influenced by over load

2. In a given space of events of plausible reasoning, two events A and B will be independent, which of the following statements is correct:

   A and B are mutually exclusive
   $P(A'\,B') = [1 - P(A)]\,[1 - P(B)]$
   $P(A) = P(B)$
   $P(A) + P(B) = 1$

3. Of the offshore structures inspected by a structural consultant in previous years, it is known that service life is influenced by a few critical factors. Out of them, it is known that 60% is due to variation (unforeseen) in environmental loads and 40% is due to material degradation. The previous year's investigations report that 30% of the offshore facilities that are influenced by overload get an "A grade" of reliability certificate and only 20% of those influenced by material degradation qualify for an "A grade" of reliability certificate. The company vets the reliability estimate made by the consultant who certifies that the structure qualifies as "A grade" of reliability. What is the probability that the structure be influenced by overloading?
4. The risk of offshore drilling rigs is being estimated on the basis of a questionnaire conducted. The engineer either knows the reason for failure from the standard questionnaire or simply guesses the reason for failure. Let 3/4 be the probability that he knows the correct reason and 1/4 be the probability that he guesses. Assuming that the engineer is experienced and capable of giving a correct guess to the probability of 1/4, what is the probability that the engineer knows the correct reason for failure given that he guessed it correctly?
5. Core cutting techniques are used to detect any deficiency in the parent concrete. In fact, such destructive tests are considered to be authentic by 99% to detect the deficiency when it is in fact actually present. However, the tests also yield a false result for 0.5% on healthy structures when conducted, that is, on a controlled concrete tested to be deficient with a probability of 0.005. With this as the background of information to ascertain the actual strength of the structure, 0.1% of the sample of structures tested was actually deficient. What is the probability that an offshore structure investigated through the test shows deficiency?
6. An insurance company insured 20 drill rig operators, 40 electricians, and 60 floor shop mechanics. Probability of accidents is 0.15, 0.03, and 0.01, respectively. One of the insured persons has an accident. What is the probability that he is a drill rig operator?
7. A factory has two machines A and B. Past record shows that machine A produced 60% of the items of output and machine produced the remaining of the items. Further, 2% of the items produced by A and 1% of that produced by B were found to be defective. All the items are put into one stock and then one item is chosen at random from the stock which was found to be defective. What is the probability that it was produced by machine B?
8. From a lot of 30 bulbs which include 6 defective bulbs, a sample of 4 bulbs is drawn at random with replacement. Find the probability distribution of the number of defective bulbs.

9. A random variable X has the following probability distribution

X	0	1	2	3	4	5	6	7
$P(X)$	0	k	$2k$	$2k$	$3k$	K^2	$2k^2$	$7k^2 + k$

a. Determine

k
$P(X < 3)$
$P(X > 6)$
$P(0 < X < 3)$

EXERCISE PAPER 5

Answer all questions

1. The random variable X has probability distribution $P(X)$ of the following form, where k is some number:

a. $$P(X) = \begin{cases} k, & \text{if } x = 0 \\ 2k, & \text{if } x = 1 \\ 3k, & \text{if } x = 2 \\ 0, & \text{otherwise} \end{cases}$$

2. Determine k
3. Find $P(X < 2)$, $P(X > 2)$, $P(X \geq 2)$
4. In an assessment report submitted by the consultant, 70% of the board members are in favor of the report and 30% disagree on certain parameters identified by the consultant as factors that influence failure. A member from the board is selected at random and it is taken that $X = 0$ if he is opposed and $X = 1$ if he is favored. Find $E(X)$ and $Var(X)$
5. There are 5% defective items in a large bulk of production. What is the probability that a sample of 10 items will include not more than one defective item?
6. The probability that a bulb produced from a factory will fuse after 150 days. What is the probability that out of five such bulbs (i) none; (ii) not more than; (iii) more than one; and (iv) at least one will fuse after 150 days of use?

EXERCISE PAPER 6

Answer all questions

1. A, B, C are waiting in a queue for a placement interview. Let us say A has half the chance of that of B to get the job and C has two-third the chance of that of B. Given that there is only one post vacant, what is the reliability of each person getting the job?
2. An instrumentation lab has three PCs that are relatively new. It is for sure that for at least one year, these PCs will work without any problem. What is the reliability that at least one PC will work after one year?

3. In a test carried out on a concrete cube for its compressive strength of design mix, it is seen that course aggregate of 30%, fine aggregate of 25%, cement f 255, and water/admixture of 20% can control the overall strength. This is arrived based on the sample tests carried out many concrete cubes with varying strength. However there is a doubt that not all samples tested are reliable. The correction/adequacy of results is represented by CA = 27%, FA = 10%, cement = 30%, and water/admixture as 50%. What is the reliability of information from the sampling at one of the random samples used for the investigation?

EXERCISE PAPER 7

PART A

Answer all questions and to the point. Each step carries credits.

1. The maximum forces on a tether of a TLP are assumed to be of Gaussian type. The Gaussian density function is given by $f_X(x;\mu,\sigma) = (1/2\pi)\exp(-(x-\mu)^2/2\sigma^2)$.

Test	Load (MN)
Case 1	500
Case 2	200
Case 3	300
Case 4	800
Case 5	560

With respect to the above information and using method of moments, write down the estimator for standard deviation. Calculate the estimate.

Can you write down another estimator for standard deviation and calculate the estimate. Comment on the results. Calculate the 90% confidence interval. (4 points)

2. Define the following with respect to a stochastic process with appropriate examples (Answer any three):
   a. Stationarity of a stochastic process
   b. Ergodic property of a stochastic process
   c. Auto correlation function
   d. Concept of ensemble averaging
   e. Monte Carlo Simulation (6 points)
3. What are the basic axioms for probability theory? Why is event space necessary in studying the probability?

   Give a distribution that best describes the offshore stochastic process with reasons:

   (a) Failure load of a steel under fatigue. (b) Marine corrosion of steel. (c) Peaks of a random process. (d) Maxima of wave loads on a structure. (e) Wave loading. (11/2 + 2 + 21/2points)

4. The statistical analysis of a stochastic process in ocean engineering is studied with respect to moments. Can you explain about the effect of non-Gaussianity with respect to skewness and kurtosis? (3 points)
5. Four time series of a force on an offshore structure are observed. The stochastic process is stationary with respect to mean for a time of 15 seconds. Justify the same, from the observed time series whether the stationarity holds good or not. One may assume an allowable tolerance of 5% among instances? (Hint: Use definition of stationarity and check only for maximum of three instances.)

	Force (MN)			
Time (s)	**Case 1**	**Case 2**	**Case 3**	**Case 4**
1.00	494.67	535.00	454.56	475.40
2.00	504.67	521.68	452.97	470.26
3.00	501.75	527.05	451.75	477.06
4.00	499.85	528.61	456.25	478.38
5.00	500.91	532.11	454.65	479.29
6.00	492.17	521.65	451.20	471.54
7.00	499.58	532.36	446.55	477.25
8.00	508.02	523.94	446.74	475.50
9.00	500.49	530.33	455.96	479.13
10.00	500.21	533.26	441.94	477.68
11.00	496.33	531.64	449.88	479.49
12.00	499.85	535.41	440.26	474.34
13.00	501.16	535.03	455.10	474.26
14.00	502.13	526.75	454.31	480.04
15.00	498.14	531.29	450.01	464.38
16.00	498.82	525.28	449.65	472.48
17.00	510.12	523.39	437.57	468.65
18.00	488.71	534.62	452.91	473.09
19.00	511.15	530.00	439.04	478.24
20.00	501.69	529.73	438.40	479.13

How does the concept of ergodic property help us in such cases? (6 points)

PART B

Answer all questions

10. What do you understand by FOSM? Derive the equation for probability of failure using FOSM. (5 points)
11. Write a brief note on the Hasofer–Lind method. (3 points)

12. Write a brief note on point-estimate-for-probability-moment methods (PEPM). Estimate the peak response of a single-degree-of-freedom system using the PEPM algorithm. (5 points)
13. In the reliability estimate of offshore structures, it is seen that 60% of the service life is influenced by overload on the structure and 40% by material degradation, if these factors are considered independently. However, on their joint contribution, their influence on the service life is about 20%. Given the above information as authentic, an offshore structure is selected at random for service life estimate. Find the probability that
    i. The service life of the structure shall not be influenced by either of these factors
    ii. If the service life is influenced by overload, then find the probability that it is also influenced by material degradation
    iii. If the service life is influenced by material degradation, then find the probability that it is also influenced by over load (3 points)
14. State different levels of reliability and briefly explain their suitability for different kinds of engineering problems. (3 points)
15. In estimating the system reliability of a given offshore structure, three independent variables namely x_1, x_2, and x_3 are identified; dependent variables namely x_4 (depends on x_1 and x_2) and x_5 (depends on x_2 and x_3) are also considered for the analysis. Using Bayesian networks and K_2 algorithm, estimate the system reliability of the platform. (6 points)

EXERCISE PAPER 8

Part A

Answer all questions and to the point. Each step carries credits.

1. List the differences between safety and reliability. (5 points)
2. What do you understand by omission sensitivity index in the reliability analysis? Derive the expression for a reliability problem with two independent, normally distributed variables with linear limit state function. (5 points)
3. Write the expression for Cornell's reliability index. Using the expression, determine the reliability index for the problem given below:
   A cantilever beam is loaded by two point loads (P_1, P_2) at distance a and $2a$ measured from the fixed end. Let the moment capacity of the beam be B. Moment representation, in second-order reliability is given by $Z = (B, P_1, P_2)$

$$E[Z] = \begin{Bmatrix} 250\,\text{kN m} \\ 10\,\text{kN} \\ 10\,\text{kN} \end{Bmatrix} \qquad C_Z = \begin{bmatrix} 900 & 0 & 0 \\ 0 & 9 & 6 \\ 0 & 6 & 9 \end{bmatrix}$$
(5 points)

4. Briefly discuss different levels of reliability, highlighting the limitations. (5 points)

5. Write a brief note on the Hasofer–Lind method (OR)
   Write a brief note on PEPM. Estimate the peak response of a single-degree-of-freedom system using the PEPM algorithm. (5 points)
6. In the reliability estimate of offshore structures, it is seen that 60% of the service life is influenced by overload on the structure and 40% by material degradation, if these factors are considered independently. However, on their joint contribution, their influence on the service life is about 20%. Given the above information as authentic, an offshore structure is selected at random for service life estimate. Find the probability that
   i. The service life of the structure shall not be influenced by either of these factors
   ii. If the service life is influenced by overload, then find the probability that it is also influenced by material degradation
   iii. If the service life is influenced by material degradation, then find the probability that it is also influenced by over load (5 points)

References

ABS. 2003. *Guide for the Fatigue Assessment of Offshore Structures*, American Bureau of Shipping, April 2003, Houston, TX.

ABS. 2004. Commentary on the *Guide for the Fatigue Assessment of Offshore Structures*, American Bureau of Shipping, Houston, TX.

Ale, BJM. 2002. Risk assessment practices in the Netherlands, *Safety Science*, 40, 105–126.

Ang, AHS, Tang, WH. 1975. *Probability Concepts in Engineering Planning and Design, Basic Principles*, Vol. 1, Wiley, New York, NY.

Ang, AHS, Tang, WH. 1984. *Probability Concepts in Engineering Planning and Design*, Vol. II. *Decision, Risk & Reliability*, John Wiley, New York.

APIRP 2T. 1997. *Recommended Practice for Planning, Designing and Constructing Tension Leg Platforms*, American Petroleum Institute, Washington, D.C.

APIRP 2A-WSD. 2000. *Recommended Practice for Planning, Designing and Constructing Fixed Offshore Platforms—Working Stress Design*, American Petroleum Institute, Washington, D.C.

API RP2-WSD. 2001. *Planning, Design and Constructing Fixed Offshore Platforms—Working Stress Design*, American Petroleum Institute, Houston, Texas.

Arshad Ayub. 2011. *Hazard Identification and Management in Oil and Gas Industries*, MTech Dissertation, IIT Madras, Chennai, India.

Aruliah, P. 1976. Fatigue durability under wave action, *Proceedings of the Institution of Civil Engineers, Part 1*, 60, 149–152.

Assakkaf, IA, Ayyub, BM. 2000. Reliability-based design for fatigue of marine structures, *Proceedings of the 8th ASCE Specialty Conference on Probabilistic Mechanics and Structural Reliability*, New York, June 2–5.

Augusti, G, Baratta, A, Casciati, F. 1984. *Probabilistic Methods in Structural Engineering*, Spon Press, Oxfordshire, UK.

AWS. 1972. *Structural Welding Code*, American Welding Society, D1, pp. 1–72.

Baltrop, NDP, Adams, AJ. 1991. *Dynamics of Fixed Marine Structures*, Butterworth Heinemann, UK.

Bathe, KJ, Wilson, EL. 1987. *Numerical Methods in Finite Element Analysis*. Prentice Hall, New Delhi.

Barlow, RE, Prroschan, F. 1981. *Statistical Theory of Reliability and Life Testing*, Silver Springs, Maryland.

Bazzurro, P, Cornell, CA. 1994a. Seismic hazard analysis of nonlinear structures, I: Methodology, *Journal of Structural Engineering*, 120(11):3320–3344.

Bazzurro. P., Cornell, CA. 1994b. Seismic hazard analysis of nonlinear structures, II: Applications, *Journal of Structural Engineering*, 120(11):3345–3365.

Benjamin, JR, Cornell, CA. 1970. *Probability, Statistics and Decisions for Civil Engineers*, John Wiley, New York.

Bhattacharyya, SK, Srinivasan Chandrasekaran, Ram Prasad. 2010a. Event analysis for offshore riser failure, *Proceedings of the 1st International Conference on Drilling Technology* (*ICDT 2010*), IIT Madras, Chennai, India, November 18–20.

Bhattacharyya, SK, Srinivasan Chandrasekaran, Ram Prasad. 2010b. Risk assessment for offshore pipelines, *Proceedings of the 1st International Conference on Drilling Technology* (*ICDT 2010*), IIT Madras, Chennai, India, November 18–20.

Bierager, A. 1990. On computational methods for structural reliability analysis, *Journal of Structural Safety*, 9(2), 79–96.

Blight, BTN, Ott, L. 1975. A Bayesian approach to model inadequacy for polynomial regression, *Biometrika*, 62, 79–88.

Bob Skelton. 1997. *Process Safety Analysis—An Introduction*, Institute of Chemical Engineers, UK.

Bonvicini, S, Leonelli, P, Spadoni, G. 1998. Risk analysis of hazardous materials transportation: Evaluating uncertainty by means of fuzzy logic, *Journal of Hazardous Materials*, 62, 59–74. Doi: 10.1016/S0304-3894 (98) 00158–7.

Borgman, LE, Hagan, RL, Kuik, AJ. 1982. Statistical precision of directional spectrum estimation with data from a tilt-and-roll buoy, *Topics in Ocean Physics*, Ed. AR. Osborne and P. Malanotte Rizzoli, North Holland Publishing Company, Amsterdam, NY, pp. 418–438.

Bottelberghs, PH. 2000. Risk analysis and safety policy developments in the Netherlands, *Journal of Hazardous Materials*, 71, 59–84. Doi: 10.1016/S0304-3894(99)00072-2.

Box, GEP, Tiao, GC. 1973. *Bayesian Inference in Statistical Analysis*, Addison-Wesley, Reading, MA.

Box, GEP, Wilson, KB. 1951. On the experimental attainment of optimum condition, *Journal of the Royal Statistical Society*, *Series B* 13, 1–34.

Bringham, EO. 1974. *The Fast Fourier Transform*. Prentice Hall, Englewood Cliffs, NJ.

Breitung, K. 1984. Asymptotic approximations for multi normal integrals, *Journal of the Engineering Mechanics Division, ASC*, 110, 357–366.

Bucher, CG, Bourgund, U. 1990. A fast and efficient response surface approach for structural reliability problems, *Structural Safety*, 7, 57–66.

Bury, KV. 1984. *Statistical Models in Applied Science*, Krieger Publishing Company, Florida.

Cartwright, DE, Longuet-Higgins, MS. 1956. The statistical distribution of the maxima of a random function, *Proceedings of the Royal Society of London, A*, 237, 212–232.

Capanoglu, CC, Shaver, CB, Hirayama, H, Sao, K. 2002. Comparison of model test results and analytical motion analyses for Buoyant Leg Structure, *Proceedings of the twelfth International Offshore and Polar Engineering Conference*, Kitakushu, Japan, May 26–31, pp. 46–53.

Copple, RW and Capanoglu, CC. 1995. A Buoyant Leg Structure for the development of marginal fields in deep water, *Proceedings of the fifth International Offshore and Polar Engineering Conference*, The Hague, The Netherlands, June 11–16, pp. 163.

CEB Bulletin. 1976. *First Order Reliability Concepts for Design Codes*, Joint Committee on Structural Safety, CEB Bulletin 112, Paris.

Chakrabarthi, SK. 2005. *Handbook of Offshore Engineering, Offshore Structure Analysis*, Inc., Vol. II. Plainfield, Illinois, USA.

Chakarabarti, SK, Hanna, SY. 1990. Added mass and damping of a TLP column model. *Proceedings of the International Offshore Technology Conference*, No. 4643, pp. 533–545.

Chandrasekaran, S, Jain, AK, 2002. Dynamic behavior of square and triangular offshore tension leg platforms under regular wave loads. *Ocean Engineering*, 29(3), 279–313.

Chandrasekaran, S, Jain, AK, 2002a. Triangular configuration tension leg platform behavior under random sea wave loads. *Ocean Engineering*, 29(15), 1895–1928.

Chandrasekaran, S, Nilanjan Saha. 2011. *Reliability of Offshore Structures*, Short Course, IIT Madras, Chennai, India.

Chen, X, Lind, NC. 1983. Fast probability integration by three parameter normal tail approximation, *Structural Safety*, 1, 269–276.

Chopra, AK. 2003. *Dynamics of Structures: Theory and Applications to Earthquake Engineering*, 2nd ed. Pearson Education, Singapore.

Coppola, A. 1984. Reliability engineering of electronic equipment: A historic perspective, *IEEE Transactions on Reliability*, R-33(1), 29–35.

Cornel, CA. 1969. A probability based structural code, *Journal of the American Concrete Institute*, 66(12), 974–985.
Crawley, F, Preston, M, Tyler, B. 2000. *HAZOP: Guide to Best Practice. Guidelines to Best Practice for the Process and Chemical Industries*. European Process of Safety Centre and Institution of Chemical Engineers, Rugby, UK.
Cronin, DJ, Godfrey, PS, Hook, PM, Wyatt, TA. 1978. Spectral fatigue analysis for offshore structures, *Numerical Methods in Offshore Engineering*, Eds. O. C. Zienkiewicz, R. W. Lewis, K. G. Stagg, Wiley-Blackwell, UK, pp. 281–316.
Crow, EL, Shimizu, K. (Eds.). 1988. *Lognormal Distributions, Theory and Applications*, Dekker, New York.
Datta, TK. 2010. Seismic reliability analysis of structures, In *Seismic Analysis of Structures*, John Wiley & Sons, Ltd, Chichester, UK. Doi: 10.1002/9780470824634.ch8.
David Brown, F, William Dunn, E. 2007. Application of a quantitative risk assessment method to emergency response planning, *Computers and Operations Research*, 34, 1243–1265.
Denson, W. 1998. History of reliability prediction, *IEEE Transactions on Reliability*, 47(3), 321–328.
Der Kiureghian, Lin, HZ, Hwang, SF. 1987. Second order reliability approximation, *Journal of Engineering Mechanics Division, ASCE*, 113(8), 1208–1225.
Ding, Y, Purath, BT, Zhang, J, Gilbert, RB, Dangyach, S, Choi, YJ. 2005. Reliability of mooring systems for a spar. *Proceedings of the 24th International Conference on Offshore Mechanics and Arctic Engineering OMAE*, Halkidiki, Greece, June 12–17, pp. 595–604.
Ditlevsen, O. 1981. *Uncertainty Modelling with Applications to Multi-Dimensional Civil Engineering Systems*, McGraw-Hill, New York.
Ditlevsen, O, Madsen, HO. 1996. *Structural Reliability Methods*, John Wiley & Sons Ltd., Chichester.
Ditlvesen, O. 1979. Generalized second—Moment reliability index, *Journal of Structural Mechanics*, 7, 435–451.
DNV. 1992. *Structural Reliability Analysis of Marine Structures*, Classification notes No. 30.6, Det Norske Veritas, Hovik, Norway.
Dowling, NE. 1972. Fatigue failure predictions for complicated stress strain histories, *Journal of Materials*, 7, 71–87.
Dziubinski, M, Fratczak, M, Markowski, AS. 2006. Aspects of risk analysis associated with major failures of fuel pipelines, *Journal of Loss Prevention in the Process Industries*, 19, 399–400.
Elishakoff, I. 1999. *Probabilistic Theory of Structures*, Dover, USA.
Engelhard, WFJM, de Klepper, FH. 1994. Hazard analysis for the Amoco Netherlands P/15-P/18 production facilities in the North Sea, *Proceedings of SPE Health, Safety and Environment in Oil and Gas Exploration and Production Conference*, 25–27 January, Jakarta, Indonesia.
Esary, JD, Proschan, F, Walkup, DW. 1967. Association of random variables with applications, *Annals of Mathematical Statistics*, 38, 1466–1474.
European Standard EN 10225:2001. Weldable structural steel for fixed offshore structures, British Standards Institution, London.
Ferry Borges, J, Castanheta, M. 1971. *Statistical Definition and Combination of Loads*, ACI Publication, SP 31, Detroit, USA, pp. 43–62.
Fiessler, B, Neumann, HJ, Rackwitz, R. 1979. Quadratic limit states in structural reliability, *Journal of Engineering Mechanics Division, ASCE*, 105(4), 661–676.
Frank, KH, Morgan, HW. 1979. A logical risk process of risk analysis, *Professional Safety*, June, 23–30.
Freudenthal, AM. 1947. The safety of structures, *Transactions of ASCE*, 112, 125–159.

Goda, Y. 1970. Numerical experiments on wave statistics with spectral simulation, *Report of the Port and Harbour Research Institute, Japan*, 9, 16–19.

Gong, SF, He, Y, Jin, WL. 2007. Fatigue life reliability analysis of single point mooring offshore jacket platform structures, *Journal of Zhejiang University*, 41(6), 995–1006.

Goodwin, P, Ahilian, RV, Kavanagh, K, Connaire, A. 1999. Integrated mooring and riser design: Reliability analysis methodology and preliminary results, *Proceedings of Offshore Technology Conference*, May 3–6, Houston, USA.

Graff, W.J. 1981. Introduction to Offshore Structures, Gulf Publishing Co., Houston.

Grandt, Jr. AF. 2004. *Fundamentals of Structural Integrity-Damage Tolerant Design and Non-Destructive Evaluation*, John Wiley, UK.

Haldar, A, Mahadevan, S. 1995. First order and second order reliability method, *Probabilistic Structural Mechanics Hand Book*, Ed. C. (Raj) Sundararajan, Chapman & Hall, London, UK, pp. 27–52.

Haldar, A., Mahadevan, S. 2000. *Probability, Reliability and Statistical Methods in Engineering Design*, John Wiley & Sons, New York.

Hasan, DN, Islam, N, Moin, K. 2009. Reliability analysis of compliant offshore tower under earthquake loads, *Proceedings of 28th International Conference on Offshore Mechanics and Artic Engineering OMAE*, Hawaii, May 31–June 5.

Haver, S, Larsed, K, Meling, TS. 2001. Reliability assessment of a mooring system, *Proceedings of the Eleventh International Offshore and Polar Engineering Conference*, 4, 490–497.

Henselwood, F, Phillips, G. 2006. A matrix based risk assessment approach for addressing linear hazards such as pipelines, *Journal of Loss Prevention in the Process Industries*, 19, 433–441.

Hovde, Olav, G, Torgeir, M. 1997. Fatigue and overload reliability of mooring systems, *Proceedings of the International Offshore and Polar Engineering Conference*, 4, 145–152.

HSE 1990. Offshore installations: Guidance on design, construction and certification, Health and Safety Executive, Bootle, UK.

IEC 61882. 2000. *Hazard and Operability Studies: Application Guide*, International Electro technical Commission, Geneva.

IS 15656. 2006. *Hazard Identification and Risk analysis*, Code of Practice, Bureau of Indian Standards, May 2006.

IS 1656. 2006, *Indian Standard Hazard Identification and Risk Analysis*, Code of Practice, Bureau of Indian Standards, 2006.

Jan Erik Vinnem. 2007. *Offshore Risk Assessment: Principles, Modeling and Applications of QRA Studies*, Springer-Verlag, London, 577pp.

John Spouge, 2005. *A Guide to Quantitative Risk Assessment for Offshore Installations*, DNV Technica, UK.

Julian, OG. 1957. Synopsis of first progress report of committee on factors of safety, *Journal of the Structural Division, ASCE*, 1316, 1–22.

Kanegaonker, HB, Belkune, RM, Ramesh, CK. 1982. On the significance of nonlinear random wave loading in the fatigue analysis of jacket type of structures, *Proceedings of the 3rd International Conference on Behaviour of Offshore Structures in Sea BOSS*, MIT, Cambridge, November 11–13.

Khan, FI, Abbasi, SA. 1999. Major accidents in process industries and analysis of causes and consequences, *Journal of Loss Prevention in Process Industries*, 12(5), 361–378.

Kim, C-H, Zou, J. 1995. A universal linear system model for kinematics and forces affected by nonlinear irregular waves, *International Journal of Offshore and Polar Engineering*, 5(3), 166–170.

Kim, C-H, Kim, MH, Liu, YH, and Zhao, CT. 1994. Time domain simulation of nonlinear response of a coupled TLP system, *International Journal of Offshore and Polar Engineering*, 4(4), 281–291.

Kinsman, B. 1965. *Wind Waves*, Prentice-Hall, NJ.

Kiran, A. 2014. Accident modeling and risk assessment of oil and gas industries, MS (by research) Thesis, IIT Madras, Chennai, India.

Kiran, A. 2012. *Risk Analyses of Offshore Drilling Rigs*, MTech Dissertation (petroleum engineering), IIT Madras, Chennai, India.

Kyriakdis, I. 2003. *HAZOP—Comprehensive Guide to HAZOP in CSIRO*, CSIRO Minerals, National Safety Council of Australia, Melbourne.

Lancaster, J. 2000. *Engineering Catastrophes*, CRC, New York.

Larsen, K, Mathisen, J. 1996. Reliability based fatigue analysis of mooring lines, *Proceedings of the 15th International Conference on Offshore Mechanic and Artic Engineering*, Florence, Italy, June 16–20, pp. 277–285.

Lin, YK. 1967. *Probabilistic Theory of Structural Dynamics*, McGraw-Hill, New York.

Lin, YK, Cai, GQ. 2004. *Probabilistic Structural Dynamics*, McGraw-Hill, New York.

Maddox, NR, Wildenstein, AW. 1975. A spectral fatigue analysis for offshore structures, *Proceedings of Offshore Technology Conference, OTC*, May 5–7, Houston, TX.

Madhavan Pillai, TM, Veena, G. 2006. Fatigue reliability analysis of fixed offshore structures: A first passage problem approach, *Journal of Zhejiang University Science A*, 7(11), 1839–1845.

Madsen, H. 1988. Omission of sensitivity factors, *Structural Safety*, 5, 35–45.

Madsen, H, Krenk, S, Lind, NC. 1986. *Methods of Structural Safety*, Prentice-Hall, NJ.

Malhotra, AK, Penzien, J. 1970. Nondeterministic analysis of offshore structures, *Journal of Engineering Mechanics Division, ASCE*, 96, 985–1003.

Marshall, PW. 1974. *Basic Considerations for Tubular Joint Design in Offshore Construction*, W.R.C. Bulletin No. 193, Welding Research Council, Cambridge, UK.

Marshall, PW. 1976. Dynamic and fatigue analysis using directional spectra, *Proceedings of Offshore Technology Conference*, OTC 1976, paper No. 2537, May 2–5, Houston, TX.

Melchers, RE. 1987. *Structural Reliability Analysis and Prediction*, Ellis Horwood Ltd, Chichester, England.

Melchers, RE. 1999. *Structural Reliability: Analysis and Prediction*, 2nd edn, John Wiley, West Sussex, England.

Mihailidou, EK, Antomiadis, KD, Assael, MJ. 2012. The 319 major industrial accidents since 1917, *International Review of Chemical Engineering*, 4(6), 1755–2035.

Michel, WH. 1999. Sea spectrum revisited, *Marine Technology*, 36(4), 211–227.

Miller, CCD. 1981. Buckling design methods for steel structures: A state of the art, *Integrity of offshore Structures*, Eds. D. Faulkner, M. J. Cowling, A. Incecik, Applied Science Publishers, Spon Press, London, pp. 397–418.

Mohd Umair. 2012. Dynamic analysis and reliability assessment of offshore spar platform, PhD Thesis, IIT Delhi, India.

Mustafa, V. 1985. Linear spectral analysis and fatigue of offshore steel jacket platforms, PhD Thesis, IIT Madras, Chennai, India.

Nigam, NC. 1983. *Introduction to Random Vibration*, M.I.T Press, Cambridge, Massachusetts.

Nigam, NC, Narayanan, S. 1994. *Applications of Random Vibrations*, Springer-Verlag, New York.

Nivolianitou, Z, Konstandinidou, M, and Michalis, C. 2006. Statistical analysis of major accidents in petrochemical industry notified to the major accident reporting system (MARS), *Journal of Hazardous Materials*, 136(1), 1–7.

Nolte, KG, Hansford, JE. 1976. Closed-form expressions for determining the fatigue damage of structures due to ocean waves, *Proceedings of Offshore Technology Conference OTC*, May 3–6, Houston.

Papazoglou, IA, Bellamy, LJ, Aneziris, ON, Ale, BJM, Post, JG, Oh, JIH. 2003. I-risk: Development of an integrated technical and management risk methodology for chemical installations, *Journal of Loss Prevention in the Process Industries*, 16, 575–591.

Papoulis, A, Pillai, SU. 1991. *Probability, Random Variables and Stochastic Processes*, 3rd edn, McGraw-Hill, New York.

Papoulis, A, Pillai, SU. 2002. *Probability, Random Variables and Stochastic Processes*, 4th edn, Tata McGraw-Hill, New Delhi, India.

Pasman, HJ, Jung, S, Prem, K, Rogers, WJ, Yang, X. 2009. Is risk analysis a useful tool for improving process safety, *Journal of Loss Prevention in the Process Industries*, 22, 769–777.

Pate-Cornell, ME. 1994. Quantitative safety goals for risk management of industrial facilities, *Structural Safety*, 13, 145–157.

Patin Stanislav. 1999. *Environmental Impact of the Offshore Oil and Gas Industry*, Eco Monitor Publishing, New York, 425pp.

Pilotto, BM, Ronalds, BF. 2003. Dynamic behavior of minimum platforms under random loads. *Proceedings of the International Conference on Offshore Mechanics and Arctic Engineering*, Cancum, Mexico, 1, pp. 441–449.

Pilotto, BM, Ronalds, BF, Stocker, R. 2002. Dynamic response of shallow water mono-pod platforms. *Proceedings of the International Conference on Offshore Mechanics and Arctic Engineering*, Oslo, Norway, 1, pp. 113–120.

Pilotto, BM, Ronalds, BF, Stocker, R. 2003. Nonlinear dynamic analysis with deterministic and random seas, *Oceans Conference Record (IEEE)*, 5, 2908–2915.

Rackwitz, R, Fiessler, B. 1976. *Note on Discrete Safety Checking When Using Non-Normal Stochastic Models for Basic Variables*, Project Working Session, Massachusetts Institute of Technology, Cambridge, MA.

Rajasankar, J, Iyer, NR, Appa Rao, TVS. 2003. Structural integrity assessment of offshore tubular joints based on reliability analysis, *International Journal of Fatigue*, 25, 609–619.

Rajashekar, MR, Ellingwood, BR. 1993. A new look at the response surface approach for reliability analysis, *Structural Safety*, 12, 205–220.

Ramamurthy, K. 2011. *Explosions and Explosion Safety*, Tata McGraw-Hill, New Delhi, India, 288pp.

Rohit Srivastav. 2015. *Experimental and Analytical Investigations on Unstiffened Tubular T and K Joints under Axial Loads and Out of Plane Bending*, MTech Dissertation, IIT Madras, Chennai, India.

Rosenblueth, E. 1975. Point estimates for probability moments, *Proceedings of the National Academy of Sciences, USA*, 72, 3812–3814.

Shah, AA, Umar, A, Siddiqui, NA. 2005. A methodology for assessing for assessing reliability of taut and slack mooring systems against instability, *Ocean Engineering*, 68, 1216–1234.

Shaver, CB, Capanoglu, CC, Serrahn, CS. 2001. Buoyant Leg Structure preliminary design, constructed cost and model test results. *Proceedings of the 11th International Offshore and Polar Engineering Conference, Stavanger*, Norway, June 17–22, pp. 432–439.

Shinozuka, M. 1983. Basic analysis of structural safety, *Journal of the Structural Division, ASCE*, 109(3), 721–740.

Srinivasan Chandrasekaran. 2010a. Chemical risks—An overview. Keynote address at HSE in oil and gas-exploration and production, *International HSE Meet, IBC-Asia*, Kuala Lumpur, Malaysia, December 6–8.

Srinivasan Chandrasekaran. 2010b. Risk assessment of offshore pipelines. Keynote address at *HSE in oil and gas-exploration and production, International HSE Meet, IBC-Asia*, Kuala Lumpur, Malaysia, December 6–8.

Srinivasan Chandrasekaran. 2011a. Strategic rig project commissioning and risk management, Keynote address at *Post Conference Workshop on International Conference on Offshore Drilling Rigs, IBC Asia*, Singapore, July 24–29.

Srinivasan Chandrasekaran. 2011b. Hazard identification and management in oil and gas industry using Hazop, *Proceedings of Seminar on Human Resource Development for Offshore and Plant Engineering (HOPE)*, Changwon University, South Korea, April, pp. 1–10.

Srinivasan Chandrasekaran. 2011c. Health, safety and environmental management in petroleum and offshore engineering, *Proceedings of Seminar on Human Resource Development for Offshore and Plant Engineering (HOPE)*, Changwon University, South Korea, April, pp. 23–28.

Srinivasan Chandrasekaran. 2011d. Quantitative risk assessment of Group Gathering Station (GGS) of oil exploration and production, *Proceedings of Seminar on Human Resource Development for Offshore and Plant Engineering (HOPE)*, Changwon University, South Korea, April, pp. 11–22.

Srinivasan Chandrasekaran. 2011e. Risk assessment and management in offshore and petroleum industries, Keynote address at *Pre-conference Workshop on International Conference on Asia Pacific HSE Forum on Oil, Gas and Petrochemicals, Fleming Gulf Conferences*, Kuala Lumpur, Malaysia, September 22–24, pp. 81.

Srinivasan Chandrasekaran. 2014a. *Advanced Theory on Offshore Plant FEED Engineering*, Changwon National University Press, Republic of South Korea, 237pp. ISBN: 978-89-969792-8-9.

Srinivasan Chandrasekaran. 2014b. Health, safety and environmental management (HSE), Keynote address at *National Research Foundation of Korea, HRD Team for Offshore Plant FEED Engineering*, Changwon National University, South Korea, February 25.

Srinivasan Chandrasekaran. 2014c. Technological advancements in process safety management, Keynote address in the *4th Annual HSE Excellence Forum in Oil, Gas and Petrochemicals*, Kuala Lumpur, Malaysia, August 19–21.

Srinivasan Chandrasekaran. 2015a. HSE in offshore and petroleum engineering, *Lecture Notes of Online Web Course, Mass Open-source Online Courses* (*MOOC*), National Program on Technology Enhancement and Learning (NPTEL), Govt. of India.

Srinivasan Chandrasekaran. 2015b. *Health, Safety and Environmental Management in Offshore and Petroleum Engineering*, John Wiley & Sons Ltd., UK, ISBN: 978-1-119-22184-5.

Srinivasan Chandrasekaran. 2015c. *Advanced Marine Structures*, CRC Press, FL, ISBN: 9781498739689.

Srinivasan Chandrasekaran. 2015d. *Dynamic Analysis and Design of Ocean Structures*, Springer, New Delhi, India, ISBN: 978-81-322-2276-7.

Srinivasan Chandrasekaran, Kiran A. 2014a. Accident modeling and risk assessment of oil and gas industries, *Proceedings of 9th Structural Engineering Convention* (*SEC 2014*), IIT Delhi, India, December 22–24.

Srinivasan Chandrasekaran, Kiran A. 2014b. Consequence analysis and risk assessment of oil and gas industries, *Proceedings of the International Conference on Safety and Reliability of Ship, Offshore and Subsea Structures*, Glasgow, UK, August 18–20.

Srinivasan Chandrasekaran, Kiran, A. 2015. Quantified risk assessment of LPG filling station, Professional safety, *Journal of American Society of Safety Engineers* (*ASSE*), September.

Srinivasan Chandrasekaran, Madhuri, S. 2015. Dynamic response of offshore triceratops: Numerical and experimental investigations, *Ocean Engineering*, 109(15), 401–409. doi:10.1016/j.oceaneng.2015.09.042.

Srinivasan Chandrasekaran, Ramesh Babu, Arshad Ayub. 2010. Hazop study for crude oil pipe line, *Proceedings of the 1st International Conference on Drilling Technology* (*ICDT 2010*), IIT Madras, Chennai, India, November 18–20.

Srinivasan Chandrasekaran, Subarata Kumar Bhattacahryya. 2012. *Analysis and Design of Offshore Structures with Illustrated Examples*, Human Resource Development Center for Offshore and Plant Engineering (HOPE Center), Changwon National University Press, Republic of Korea, 285pp, ISBN: 978-89-963915-5-5.

Technip. 1981. Wave and wind directionality: Applications to design of structures, *Proceedings of the International Conference on Wave and Wind Directionality*, Paris, September 21–25.

Terje Aven, Jan Erik Vinnem. 2007. *Risk Management with Applications from Offshore Petroleum Industry*, Springer-Verlag, London, 200pp.

Throft-Christensen, P, Baker, M. 1982. *Structural Reliability Theory and Applications*, Springer-Verlag, Berlin.

Thoft-Christensen, P, Dalsgard Sorensen, J. 1982. Reliability of structural systems with correlated elements, *Applied Mathematical Modelling*, 6(3), 171–178.

Trevor Kletz. 2003. *Still Going Wrong: Case Histories and Plant Disasters*, Elsevier, UK, 230pp.

Tromans, PS, Hagemeijer, PM, Wassink, HR. 1992. The statistics of the extreme response of offshore structures, *Ocean Engineering*, 19(2), 161–181.

Tvedt, L.1990. Distribution of quadratic forms in normal space—Application to structural reliability, *Journal of the Engineering Mechanics Division, ASCE*, 116(6), 1183–1197.

Vendhan, CP. 2004. The impact of ocean wave dynamics on ocean engineering, *Proceedings of the Indian National Science Academy*, 70(6), 881–891.

Venkata Kiran, G. 2011. *QRA in Oil and Gas Industries using PHAST RISK*, MTech Dissertation (petroleum engineering), IIT Madras, Chennai, India, submitted.

White, NC, Copple, WR, Capanoglu, C. 2005. Triceratops: An effective platform for developing oil and gas fields in deep and ultra-deep water, *Proceedings of the fifteenth International Offshore and Polar Engineering Conference*, Seoul, Korea, June 19–24, pp. 133–139.

William J. Cairns (Ed.), 1992. *North Sea Oil and the Environment: Development Oil and Gas Resources, Environmental Impacts and Responses*, International Council of Oil and the Environment, CRC Press, Florida, p. 712, ISBN-1: 1851667040.

Williams, AK, Rinne, JE. 1976. Fatigue analysis of steel offshore structures, *Proceedings of the Institution of Civil Engineers, Part 1*, 60, 635–654.

Wirsching, PH. 1984. Fatigue reliability for offshore structures, *Journal of Structural Engineering, ASCE*, 110(10), 2340–2356.

Wirsching, PH, Light, MC. 1980. Fatigue under wide band random stress, *Journal of Structural Engineering, ASCE*, 106, 1593–1607.

Wirsching, PH, Paez, TL, Ortiz, K. 2006. *Random Vibrations: Theory and Practice*, Dover, USA.

Young-Do Jo, Bum Jong Ahn. 2005. A method of quantitative risk assessment for transmission pipeline carrying natural gas, *Journal of Hazardous Materials*, A123, 1–12.

Index

A

Acceptable risk, 207, 208
Accident loads, 37
ALARP, *see* As low as reasonably practical (ALARP)
Aleatory type, 59
Algorithmic uncertainties, 3
As low as reasonably practical (ALARP), 62, 177, 207
Associativity equation, 15
Asymptotic approximations, 75
Auger's operation condition, 104, 105
Autocorrelation function, 137
Autocovariance function, 137
Automatic blocking system, 211

B

Barrier crossing, 139
Bayesian approach, 5
    full Bayesian approach, 6
    modular Bayesian approach, 5–6
Bayesian class, 9
Bayesian rule, 66
Bayes' theorem, 6, 8
Bernoulli sequence, 29
Bernoulli variables, 83
Bessel's function, 98, 100
BLEVE, *see* Boiling liquid expanding vapor explosion (BLEVE)
Blocking systems, 211
BLS, *see* Buoyant leg structures (BLS)
Boiling liquid expanding vapor explosion (BLEVE), 205, 213
Boolean algebra, 14, 20
Boolean variables, 124, 125
Bund, 212
Buoyant leg structures (BLS), 133, 220

C

Cause–consequence analysis, 203–204
cdf, *see* Cumulative distribution function (cdf)
Chebyshev coefficients, 78
Chebyshev polynomial, 78
Checking point, 72
Chi-square distribution, 35
Chord saddle point, 161
Classical Sturm–Liouville problem, 99
Coefficient of inertia, 113
Coefficient of variation (COV), 30, 131
Complete beta function, 22
Composite
    exposure, 193–194
    sampling density, 36
Compound hypotheses, 22–23
Conditional distributions, 28
Conditional probability, 19, 69
Consequence analysis, 205–206, 208
Consequential effects, 205–206
Consequential errors, 65
Continuous probability distribution functions, 19–20
Corrective actions, 199
COV, *see* Coefficient of variation (COV)
Cumulative damage estimation, 144
Cumulative density function, *see* Cumulative distribution function (cdf)
Cumulative distribution function (cdf), 20, 26, 87, 120
Cut sets, 80–81

D

Damage criteria selection, 214–215
Damping matrix, 109
Decision analysis, 204
Decision trees, 204–205
Deductive logic, 9
    logical reasoning, 13
    product and sum rule, 11
    propositions, 10
    qualitative rules, 11
    Total probability theorem, 12
Deductive reasoning, 13
    qualitative rules, 17–19
    quantitative rules, 14–17
Deformation loads, 37
de Morgan's laws, 87
Design point, *see* Probable failure points
Detection, 198
Dimension reduction (DR), 4
Dirac-delta function, 139
Directional simulation, 34–36
Discrepancy function, 5
Discrete wave approach, 140–141
    fatigue assessment using, 145
    fatigue damage, 146–148
    S–H relationship, 145–146
Disjoint cut sets, 81
Disjoint path sets, 81

Distribution parameter estimation, 38
least squares, 39
maximum likelihood, 39
method of moments, 39
probability plots, 40
DR, *see* Dimension reduction (DR)
Dynamic amplification effect, 148–149
Dynamic buoyant force, 110
Dynamic tether tension, 110, 113

E

Emergency shut down system (ESD system), 211
Empirical rules, 67
Endurance limit, 142
Environmental loads, 1
modeling, 37
offshore structures, 37
return period, 37–38
Epistemic type, 59
Equally correlated elements, system with, 120
limit state equation, 121
reliability index, 121, 122
Error estimation, 64
approaches, 65
classification of errors, 65
Errors of concept, 65
Errors of execution, 65
Errors of intention, 65
ESD system, *see* Emergency shut down system (ESD system)
ETA, *see* Event tree analysis (ETA)
Event tree, 199
Event tree analysis (ETA), 178, 203
Evidence theory, 4
Excess flow valve, 212
Expectation operator, 30
Experimental uncertainties, 4
Explosion, 215
Extreme value distributions, 36

F

Fabrication errors, 67
Failure domain, 60
for general system, 85
for parallel system, 84
of series system, 84
Failure mode and effect analysis (FMEA), 178, 195
applications, 197–198
methodology, 195–197
road map, 198
variables, 198–199
worksheet, modified with details, 200
Failures, 59
Failure scenario selection, 210
blocking systems, 211
bund, 212
excess flow valve, 212
intervention by operators, 212–213
LOC, 210
non-return valve, 212
total release duration, 211
Fatigue, 140
damage, 140–143, 146–148
spectrum, 144
Fatigue assessment
using discrete wave approach, 145
fatigue damage, 146–148
S–H relationship, 145–146
simplified fatigue assessment method, 148–149
Fatigue design factors (FDF), 141
Fatigue limit states (FLS), 126
Fatigue reliability, 140–141
lognormal format for, 153–154
uncertainties in, 152–153
Fault tree analysis (FTA), 178, 201
block diagram for system reliability, 202
combination of logical gates, 202
rules of probability theory, 202
symbols in, 201
system failure modes, 203
Fault tree, 199
FDF, *see* Fatigue design factors (FDF)
FEED, *see* Front end engineering design (FEED)
FEM, *see* Finite element model (FEM)
FFNI, *see* Full factorial numerical integration (FFNI)
Finite element model (FEM), 67
Fire radiation, 213–214
First-order estimates, 86
behavior of general system, 88
de Morgan's laws, 87
inclusion–exclusion of set theory, 88
inequality event, 90
probability of failure, 86
system failure domain of first-order approximation, 89
First-order reliability method (FORM), 4
First-order second moment method (FOSM method), 69–71; *see also* Hasofer–Lind method
First level of risk assessment, 189
FLS, *see* Fatigue limit states (FLS)
FMEA, *see* Failure mode and effect analysis (FMEA)
"*F–N* pair", 217
FORM, *see* First-order reliability method (FORM)
Forward uncertainty propagation, 4–5
FOSM method, *see* First-order second moment method (FOSM method)

Frequency
analysis, 208
class, 9
Front end engineering design (FEED), 179
FTA, *see* Fault tree analysis (FTA)
Full Bayesian approach, 6
Full factorial numerical integration (FFNI), 4
Functional expansion-based methods, 4
Fuzzy theory, 4

G

Galerkin's variation method, 100
Gamma functions, 147, 152
Gas–oil ratio (GOR), 181
Gaussian density plot, 128
Gaussian distribution, 128, 149
Gaussian process (GP), 5, 138
General systems, 80
cut sets, 80–81
failure domain for, 85
with minimum cut sets, 82
minimum path sets, 82
path sets, 81
system functions for, 81–83
Geometric variables treatment, 126
GOR, *see* Gas–oil ratio (GOR)
GP, *see* Gaussian process (GP)
Gravity loads, 37
Grouping, 63

H

Hand-operated blocking system, 211
Hasofer–Lind method, 71
algorithm, 74–75
mean values and standard deviations of equivalent normal variables, 74
minimum distance, 72, 73
reduced variables, 71, 72
reliability index, 72–74
Hazard analysis, 178
Hazard and operability study (HaZop study), 178
advantages and limitations, 187–188
applicability, 179–180
causes, 183
consequences, 184
deviations, 183
guide words, 182
parameter and guide word combinations for, 183
parameters for, 181
record sheet, 186
safeguards, 184–187
study process, 180–181
Hazard assessment, 208–209
Hazard identification, 208–210
HaZop study, *see* Hazard and operability study (HaZop study)
Heat radiation, 214
Heave responses, 115
High-order stochastic response surface method, 77–78
Higher-order response surface methods, reliability estimate using, 76–77
Hinged joints geometric modeling, complexities in, 133
connecting deck and buoyant legs, 134
offshore triceratops, 134
stiffness and damping, 133–134, 135
Hot-spot stress, 157
Human error, 67
Human failure, 184–185
Hydrocarbon plants, 182
Hypergeometric distribution, 24
Hyper parameters of GP model, 5

I

Ideal sampling function, 31
Inclusion–exclusion of set theory, 88
Incomplete gamma function, 147
Index set, 136
Indicator function, 29, 79
Individual risk, 208, 216–217
Individual risk per annum (IRPA), 217
Internally ring-stiffened joint, 156
Interpolation uncertainties, 4
Interval analysis, 4
IRPA, *see* Individual risk per annum (IRPA)
Iterative response surface approach, 77

J

Jet fire, 213
Joint distribution function, 137

K

k-out-of-*n* system, 80
Kanai–Tajimi model, 111
ground acceleration spectrum, 106
Karhunen–Loeve expansions (KLE), 4
K joint(s); *see also* T joint(s)
allowable number of cycles, 175
under axial loading, 171, 173
fatigue life, 176
SCF equation and parameter validation, 174
SCF value, 172
KLE, *see* Karhunen–Loeve expansions (KLE)

L

Latent failure, 185
Least squares, 39
Level I method, 63, 64
Level II method, 63
Level III method, 63
Level IV method, 63, 64
Likelihood, 184
Limit state functions, 61, 68, 70, 72, 75, 77, 126
Linear damage hypothesis, *see* Cumulative damage estimation
Live loads, 37
Load and resistance factor design method (LRFD method), 63, 127
Load treatment, 126
LOC, *see* Loss of containment (LOC)
Local expansion-based methods, 4
Location-specific individual risk (LSIR), *see* Individual risk
Logical risk analysis, 188
  example problem for risk analysis, 194
  relative risk, 193–194
  risk index, 189–192
Loss of containment (LOC), 210
LRFD method, *see* Load and resistance factor design method (LRFD method)

M

Markov chain Monte Carlo (MCMC), 6
Material properties treatment, 126
Material safety data sheet (MSDS), 187
Mathieu stability analysis, 98
  Auger's operation condition, 104, 105
  Bessel's function, 98, 100
  dynamic equation for tether's free lateral vibration, 99
  Galerkin's variation method, 100
  generalized displacement amplitudes, 103–104
  geometric properties of tension leg for stability analysis, 102
  Mathieu's stability chart, 102
  plan of TLP, 101
  stability parameters for TLP_1, 102
  tether's dynamics, 101
Maximum likelihood, 39
Maximum likelihood estimate (MLE), 5, 6, 39
MCMC, *see* Markov chain Monte Carlo (MCMC)
Mean sea level (MSL), 106
Mechanical model, 131–132
  complexities in hinged joints geometric modeling, 133–135
  complexity, 132
Mechanical modeling, 129–130
Method of moments, 39
Miner rule, 149
Minimum cut set, 81, 82, 203
Minimum distance, 72, 73
Minimum path set, 81, 82
MLE, *see* Maximum likelihood estimate (MLE)
Model uncertainty, 66
Modified Morison's equation, 110
Modular Bayesian approach, 5–6
Monte Carlo method, 131
  simulation method, 28–30
Morgan hazard checklist, 192
Morgan's method, 194
Most probable point-based methods (MPP-based methods), 4
MPP-based methods, *see* Most probable point-based methods (MPP-based methods)
MSDS, *see* Material safety data sheet (MSDS)
MSL, *see* Mean sea level (MSL)
Multiple random variables, 27

N

Nataf random variables, 27–28
Newmark's β time integration procedure, 110
Node identification, 180
Non-return valve, 212
Nonlinear least squares, 39
Normalizing constant, 21
Norwegian Standards (NS3472E), 145
NS3472E, *see* Norwegian Standards (NS3472E)
Nuisance parameter, 22, 23
Numerical integration methods, 3
  integration-based methods, 4

O

Occurrence, 198
Off-diagonal terms in correlation matrix, 87
Offshore structures, 37
  uncertainties in system design of, 67
Offshore triceratops, 134, 219
  component breakdown and failure modes, 221
  FMEA-offshore triceratops, 223–224
  FMEA methodology for, 220
  oil exploration, 220
  risk analysis, 225
  risk analysis summary for, 222
  triceratops, 221
One-dimensional Chebyshev polynomial, 78
Out-of-plane bending, 166
  fatigue life, 171
  hot-spot stress, 167, 168
  SCF comparison at T joint, 169, 170
  T joint under, 166, 167

P

P&ID, *see* Piping and instrumentation diagram (P&ID)
Palmgren–Miner's rule, 144, 146, 153
Parallel system, 80
Parameter uncertainties, 3
Parametric variability, 3
Partial coefficient evaluations, 127
Partial safety factors, 63
Patent failure, 185
Path sets, 81, 83
PCE, *see* Polynomial chaos expansion (PCE)
pdf, *see* Probability density function (pdf)
Peak distribution, 139–140
Performance function, 71
Periodic underwater inspection, 67
Permanent loads, 37
Person-specific individual risk, 216
PHA, *see* Preliminary hazard analysis (PHA)
Pierson–Moskowitz spectrum (PM spectrum), 106, 108
Piping and instrumentation diagram (P&ID), 179, 185
Plausible reasoning principles, 8–9
PLL, *see* Potential loss of life (PLL)
PM spectrum, *see* Pierson–Moskowitz spectrum (PM spectrum)
Polynomial chaos expansion (PCE), 4
Polynomial order, 78
Pool fire, 213
Possibility theory, 4
Post-loss objectives, 218
Potential loss of life (PLL), 217
Power spectral density (PSD), 115
Practical reliability, 60
Pre-loss objectives, 218
Pre-posterior analysis, 205
Preliminary hazard analysis (PHA), 178
Pressure, 181
Prior decision analysis, 205
Probabilistic approach, 62–63
    for uncertainty propagation, 4
Probabilistic/probability, 2
    analysis, 9
    of failure, 62, 63, 68, 70, 86, 87
    modeling, 62, 66
    plots, 40
    rules of, 7–8
Probability density function (pdf), 20, 120
Probability distributions, 38
    function, 20
Probable failure points, 72, 132
Product rule, 11, 14, 17
    associativity equation, 15
    Boolean algebra, 14
    monotonic function, 16
Propositions, 10
PSD, *see* Power spectral density (PSD)

Q

QC, *see* Quality certification (QC)
Qualitative risk assessment (QRA), 178
    limitations, 206
Qualitative rules, 17–19
Quality assurance, 65
Quality certification (QC), 41
Quantitative rules, 14
    product rule, 14–17
    sum rule, 17

R

Randomness and variability of environmental loads, 66
Random number, 26
    generation with specified distribution, 26–27
Random variables, 26, 61; *see also* Variables
    conditional distributions, 28
    multiple random variables, 27
    Nataf random variables, 27–28
    plot of variables, 27
    random numbers generation with specified distribution, 26–27
RAO, *see* Response amplitude operators (RAO)
Realization, 136
Reduced variables, 71, 72
Relative risk, 193–194
Reliability, 2, 59, 66; *see also* System reliability
    assurance, 65
    bounds for structural systems, 124–125
    components of reliability analysis, 60
    estimate using higher-order response surface methods, 76–77
    against failure, 68
    failures, 59
    index, 62, 63, 70
    levels, 63–64
    limit-state function, 61
    methods, 1, 69
    practical and theoretical, 60
    problem, 68–69
    uncertainties, 59
Reliability analysis, 132; *see also* K joint(s); T joint(s)
    barrier crossing, 139
    characteristic value of basic variables, 126–127
    cumulative damage estimation, 144
    fatigue assessment, 145–148
    fatigue damage, 141–143

Reliability analysis (*Continued*)
fatigue reliability, 140–141
fundamental analysis, 120
Gaussian process, 138
limit state functions, 126
lognormal format for fatigue reliability, 153–154
mechanical modeling, 129–135
peak distribution, 139–140
reliability bounds for structural systems, 124–125
reliability coupling, 131–132, 132–135
reliability sensitivity analyses, 131
risk, 119
short-term fatigue damage, 150–152
simplified fatigue assessment method, 148–149
S–N curves, 141–145
spectral fatigue analysis of offshore structures, 149–150
stochastic modeling, 127–129
stochastic process, 135–138
structural codes application on safety, 125
system with equally correlated elements, 120–122
system with unequal correlated elements, 122–123
tubular joints, 154–157
uncertainties in fatigue reliability, 152–153
Reliability coupling, 131–132
complexities in hinged joints geometric modeling, 133–135
complexity, 132
Reliability of system (*Rsys*), 36
Remote-controlled blocking system, 211
Repression systems, 211
Response amplitude operators (RAO), 221
Response surface method, 77
Return period, 37–38
Risk, 2, 119, 216
reduction, 209
societal, 208, 217–218
Risk acceptance criteria, 207–208
Risk analysis of structures, 66
Risk assessment, 209
cause–consequence analysis, 203–204
consequence analysis, 205–206
damage criteria selection, 214–215
decision trees, 204–205
event tree analysis, 203
failure scenario selection, 210–213
fault tree analysis, 201–203
fault tree and event tree, 199
fire and thermal radiation, 213–214
FMEA, 195–199
and hazard assessment, 208–209
hazard identification, 178, 209–210
individual risk, 216–217
limitations of QRA, 206
logical risk analysis, 188–194
and management, 218
matrix, 219
objectives of risk management, 218–219
offshore triceratops, 219, 220–225
oil and gas installations, 177
quantification, 178
risk acceptance criteria, 207–208
risk picture, 216
societal risk, 217–218
Risk index, 189
first level of risk assessment, 189
Morgan hazard checklist, 192
parameters for risk analysis, 190–192
Risk management, 218
objectives, 218–219
Risk priority number (RPN), 178
Rosenblatt transformations, 28
RPN, *see* Risk priority number (RPN)
Rules of probability, 7–8
Rules of thinking, 8

S

Safety, 65
Safety domain, 60
domain for general system, 85
Sampling
advantages, 34
expectation operator, 30
over half-space, 31
within hyper-rectangle, 31
ideal sampling function, 31
importance, 30
improvement of failure estimate in two-dimensional domain, 32
with normal density at design point, 33
typical case of failure domain, 33
SCFs, *see* Stress concentration factors (SCFs)
Second-order reliability method (SORM), 4, 75
Series system, 79
Serviceability limit states (SLS), 126
Severity, 198
Shape parameter, 40
Short-term fatigue damage, 150
evaluation of damage integrals, 151–152
stress spectrum, 150
Wirsching rain flow factor, 150
S–H relationship, 145–146
Simple hypotheses, 22–23
Simulation-based methods, 4
reliability method, 76
SLS, *see* Serviceability limit states (SLS)
S–N curve, 141–143
designing, 144–145

Societal risk, 208, 217–218
SORM, *see* Second-order reliability method (SORM)
Source-strength parameters, 205
Space
  of events, 7
  of variables, 64
Specific conditions, 2
Stability of tethers under distinctly high sea waves and seismic excitation, 105
  assumptions, 106
  coefficient for interpolation, 113
  coefficient of inertia, 113
  damping matrix, 109
  dynamic buoyant force, 110
  frequency domain response of TLPs, 116
  frequency responses, 115
  geometrical properties of example TLP, 112
  Kanai–Tajimi model, 111
  natural wave periods and natural frequencies of triangular TLPs, 113
  phase plots of response of TLPs, 117
  PM spectrum, 106, 108
  PSD of vertical ground acceleration, 111
  response summary of triangular TLPs, 114
  response time histories of triangular TLPs, 115
  time history of ground velocity, 112
  TLP, 109
  triangular TLP model, 107
  wave elevation, 107
Standard Gram–Schmidt algorithm, 75
Standard normal variables, 87
Statistical theories of extremes, 36
Statistical uncertainty, 66
Stochastic modeling, 127
  Gaussian density plot, 128
  Gaussian distribution, 128
  representations of variables, 127
  statistical methods, 127
  wind speed, 129
Stochastic process, 135
  joint distribution function, 137
  realization, 136
  stochastic variables, 136
  strictly homogeneous process, 137–138
Stress concentration factors (SCFs), 145, 155
Strictly homogeneous process, 137–138
Structural reliability theory, 59
  application problems, 98–117
  computing system reliability, 83–85
  error estimation, 64–65
  first-order estimates, 86–90
  FOSM method, 69–71
  general systems, 80–81
  Hasofer–Lind method, 71–75
  high-order stochastic response surface method, 77–78
  probabilistic approach, 62–63
  reliability, 59–61
  reliability and quality assurance, 65
  reliability estimate using higher-order response surface methods, 76–77
  reliability levels, 63–64
  reliability methods, 69
  reliability problem, 68–69
  second-order reliability methods, 75
  simulation-based reliability method, 76
  space of variables, 64
  system functions for general systems, 81–83
  system reliability, 78–80
  tutorials and solutions, 90–98
  uncertainties inherent in design, 65–67
  uncertainties in system design of offshore structures, 67
  variables in reliability study, 61–62
Structural systems, reliability bounds for, 124–125
Structural uncertainties, 3
Subjective probability, 7
Sum rule, 11, 17
System failure modes, 203
System function, 124
  for general systems, 81–83
System reliability, 78
  computing, 83–85
  k-out-of-*n* system, 80
  parallel system, 80
  series system, 79
  variable, 79
System with equally correlated elements, 120
  limit state equation, 121
  reliability index, 121, 122
System with unequal correlated elements, 122–123

T

Taylor series expansion of performance function, 71
Temporary failures, 59
Tension leg platforms (TLPs), 61, 109
  frequency domain response of, 116
  generalized displacement amplitudes, 103–104
  geometrical properties of example, 112
  geometric properties of tension leg for stability analysis, 102
  natural wave periods and natural frequencies of triangular TLPs, 113
  phase plots of response of, 117
  plan of, 101
  response time histories of triangular TLPs, 115
  stability parameters, 102
  triangular TLP model, 107

Testing of hypotheses, 20–22
Tether's dynamics, 101
Theoretical reliability, 60
Thermal radiation, 213–214
Time-invariant problems, 68
Time-variant problems, 68
Time-variant reliability problem, 69
T joint(s); *see also* K joint(s)
  A-frame, 157–158
  API RP 2A recommendations, 160–161
  behavior under axial loads, 157, 159
  Chord saddle point, 161
  failed T joint, 165
  under out-of-plane bending, 166–171
  ovalization of chord, 164
  parameters, 160
  Poisson's ratio, 159
  SCF comparison at T joint, 163
  SCF variation, 161, 162, 166
  strains, 158–159
TLPs, *see* Tension leg platforms (TLPs)
Total Probability Theorem, 8, 12
Triangular TLP model, 107
  natural wave periods and natural frequencies of, 113
  response summary of, 114
  response time histories of, 115
Tubular joints, 154; *see also* K joint(s); T joint(s)
  fatigue life estimate, 156–157
  internally ring-stiffened joint, 156
  nomenclature of joint, 155
  SCF, 155
  unstiffened joint, 155–156

U

Ultimate limit state (ULS), 126
Uncertainties, 2, 59
  algorithmic, 3
  experimental, 4
  inherent in design, 65–67
  interpolation, 4
  parameter, 3
  parametric variability, 3
  structural, 3
  in system design of offshore structures, 67
Under universal testing machine (UTM), 41
Unequal correlated elements, system with, 122–123
Unit direction vector, 34
Urn distribution, 23
  using Bayes theorem, 23
  expected value, 26
  hypergeometric distribution, 24
  posterior probability distribution, 24, 25
UTM, *see* Under universal testing machine (UTM)

V

Vapor cloud explosion (VCE), 213–214
Variables, 79; *see also* Random variables
  Bernoulli, 83
  Boolean, 124, 125
  characteristic value, 126–127
  FMEA, 198–199
  interrelationship of, 62
  reduced, 71, 72
  in reliability study, 61
  space of, 64
  standard normal, 87
  structural engineer, 61
VCE, *see* Vapor cloud explosion (VCE)

W

Water particle kinematics, 110
Wave elevation, 107
Weak link, 197
Weakly homogeneous process, 138
Weakly stationary process, *see* Weakly homogeneous process
Weibull distribution, 148
  function, 152
Wirsching rain flow factor, 150

